DAIMLER
&
LANCHESTER

A Century of Motoring History

by

Brian Long

FOREWORD

It was with much delight and interest that I read Brian Long's work "Daimler & Lanchester - A Century of Motoring History" It is a fascinating story, full of interesting facts, details and photographs, which should appeal to a wide circle of Daimler enthusiasts, motorists, engineers and historians alike.

When invited to write a Foreword for this book, it brought back many memories of my time as an apprentice in the Experimental Department before the war (under Bob Crouch), as well as my time after the conflict in the Drawing Office, and later as Chief Engineer in Experimental.

Some of the early Daimler history, pointed out to me by the author, appears in print for the first time, and there are some facts that St John Nixon was either not allowed to mention, or not allowed to see - his book being commissioned by the Company!

Mr Long has admirably charted the problems and highlights of the Company's fortunes in the post-war years prior to the Jaguar takeover in 1960. I then moved to Daimler Bus Division.

I hope you will enjoy this volume as much as I did, it is indeed a fitting tribute to the Daimler and Lanchester marques.

R.R. 'George' Fabel
Kenilworth, 1995

First published in 1995 by Longford International Publications, Great Britain.
Distributed by Longford International. Fax: 01926-812327.

ISBN 1 899154019

British Library Cataloguing in Publication Data.
A catalogue record for this book is available from the British Library.

Typesetting and design by Longford Publications.
Printed in England by Warwick Printing Company Ltd.

DAIMLER
&
LANCHESTER

A Century of Motoring History

CONTENTS

ACKNOWLEDGEMENTS

Over the years I have spent researching for this book, I have come into contact with many people who have helped me with the project. It would be impossible to list everyone, but I feel I must thank the following individuals and organisations.

The help given to me by members of staff at Jaguar Cars Limited has been nothing short of outstanding. In particular, however, I should like to thank the entire Public Relations staff, Roger Clinkscales, Keith Cambage (and Sandra), Ian Luckitt, Gavin Thompson (and Sue), Joe Greenwell, Val West, Ken Edwards, Peter Banks, John Vernon, Geoff Lawson, Ann Harris, Tony O'Keefe, Sue Pearson and Jim Randle.

Members of the Daimler & Lanchester Owners Club have also helped enormously. My thanks to Maurice Ford, David Adcock, George Stapley, Tony Freeman, Tony Bagley, Simon Pickford, Glenn Westwood, John Ridley, John Box, Alan Davis, Ralph Cartwright, John Wentworth, Chris Wiltshire, Reg Turner, Mike Riley, Russ Carpenter, Ernest Leyland, Keith Sutherby, John Pochin, Duncan Saunders, Chris Hood and Dave Newman. Special thanks are due to Brian Smith for his kind co-operation.

A considerable amount of help has been received from Eric Shadforth at Unipart, Terry Phillips of Park Sheet Metal, David Watson and Bill Gettings at Motor Panels, the staff from Avon Coachworks, Mike McDowell of Coombs, Ken Bilson at Startins, and Phillip Smillie of Stratstones. Thanks to the numerous individuals, such as Colin Bromfield, 'George' Fabel, William Winters, Rupert Instone, the late Andrew Whyte, Paul Skilleter, 'Lofty' England, Micheal Colgrave, Maurice Degonda, Alec Wade, Ivor Bleaney, Ian Mobley, Reg Bowdler, Phil Weaver and the many helpful ex-Daimler workers I met on my travels. Sincere thanks also to the Royal Mews for allowing me to go into such detail on the Royal Daimlers.

A great deal of assistance was given to me by the Museum of British Road Transport in Coventry, both in the way of photographs and information. Thanks must go to Steve Bagley, Barry Littlewood, and Chris Boyce. The National Motor Museum at Beaulieu were, as ever, most helpful, as were the staff at the Patrick Collection and the Science Museum in London. Mercedes-Benz AG, the Midland Air Museum, and Quadrant Picture Library also helped.

I am greatly indebted to the City Records Office in Coventry, the City's Central Library (in particular the Local Studies section), Richard Storey at Warwick University, Companies House, and the Herbert Art Gallery & Museum. The new BMIHT facility at Gaydon has also recently proved to be a great asset to the study of British motoring.

Lastly, I would like to acknowledge the Autocar and Motor magazines, Classic Cars, Classic & Sportscar, MotorSport, Old Car, Jaguar Quarterly (now Jaguar World), The Driving Member, Jaguar Driver, Legend, and Veteran & Vintage, as they have all proved to be an invaluable source of information. The books, Daimler 1896-1946, Forty Years of Progress, The Daimler Tradition, Royal Daimlers and Jaguar; the definitive history, were constantly referred to, as were a number of others, sadly far too many to mention.

Without doubt, I will have failed to mention many names, so please accept my apologies. Thank you to everyone who has encouraged me to persevere with this project, especially my long-suffering parents and my wife, Miho. Finally, many thanks to the publishers, and naturally, to George Fabel for writing the Foreword.

Brian Long
Coventry, 1995

PREFACE

Daimler is one of those names in the motoring world which conjures an image of large stately limousines taking VIP's from one majestic setting to another. However, Daimler, with its origins dating from the late-nineteenth century (when the horseless carriage was still in its infancy on the Continent, let alone here in the British Isles), has played an important role in the history and development of the motor car. In 1897, the *Hardware Trade Journal* observed:

"The time seems approaching when there will be produced automotives, or, as they are often called, motor-cars, which will be of practical use, and which may expect to receive a fair measure of public favour. That automobiles will ultimately become ordinary vehicles for pleasure and trade purposes is certain..."

The Daimler Motor Company was instrumental in making this statement a reality, however far-fetched it must have seemed at the time. Although the cars went on to be recognised the world over for their high quality and reliability, Daimler's massive contribution to the motor industry often seems to be forgotten. Writing in 1961, William Boddy, the long-serving Editor of *MotorSport* magazine summed up the marque by stating: *"All Daimlers were dignified, rather complex, and typically English specimens of automobile engineering."*

Lanchester, on the other hand, is a make rarely given the credit it is due. As the builder of Britain's first four wheeled motor car, and one of the most innovative of all manufacturers, one would think more people would acknowledge their achievements - many features invented by Dr Fred Lanchester were found in cars decades after he had first thought of them, and some can still be found in models from the present day. In 1931 however, weakened by the Depression, they were taken over by Daimler, which is where they fit into this story.

After the Second World War, Daimler continued to produce prestige vehicles of high quality, setting them apart from the norm. The early post-war years were exciting times at the Coventry factory until politics turned things sour in the mid-1950s. Few people will forget the heady days of the Docker years, which came to a sudden end in 1956.

Shortly after this, Jaguar took the business over, by which time, the Lanchester marque had joined the growing list of names lost to history. It was a worry for a time that Jaguar would not continue to market Daimler cars, but fortunately they did. Andrew Whyte, one of the world's finest motoring historians, while in Jaguar's Public Relations Department, wrote:

"It is a gratifying thought that the name 'Daimler' - one of the greatest in motoring history - should be perpetuated, particularly as so many of the other great marques of pioneer days no longer exist."

In more recent times, badge engineered Jaguars have played their part in keeping the Daimler tradition alive. Tony Freeman noted in his book published in 1990:

"There is something oddly eccentric about a people who can draw distinctions in social status from a mere badge, but it still happens. How many British car parks still have spaces for the Directors' Jaguars and the Chairman's Daimler? The very existance of the marque over the last 30 years has depended on this quirk, which is some indication of the goodwill attached to the name."

Many other great concerns connected with Daimler are covered, so that firms like Lanchester, BSA, Hooper, Barker, Vanden Plas, and even Jaguar get an airing. The story of these colourful companies is traced, from their formation, right the way through a century of history. It is hoped this book will serve as a tribute to the many pioneers who made it all possible, and to those who continue to keep the legend alive.

CHAPTER ONE

THE FORMATIVE YEARS

The long and eventful story of the British Daimler begins in the year 1890, with an exhibition devoted mainly to modern engineering at Bremen in northern Germany. It was here that a young engineer by the name of Frederick Richard Simms displayed his aerial cableway which he had designed and built in conjunction with a friend.

A narrow gauge passenger trolley car was also shown at the exhibition. It had already proved itself admirably on a small stretch of tramway between the German towns of Unterboihinger and Kircheim-unter-Teck during 1887, but by far the most interesting feature was its means of propulsion; a small petrol powered vertical internal combustion engine.

Its high running speed, its compactness and remarkable power output for such a unit, impressed every engineer who examined it, but none more so than Simms. Simms was quick to realise the engine's great possibilities for use as a motive power unit with numerous applications on both land and water. He lost no time in making contact with the engine's inventor, Gottlieb Daimler, along with his partner, Wilhelm Maybach. From the very first meeting, a feeling of mutual trust and respect was established between the two men, even though Daimler was nearly twenty years Simms' senior.

Facing page:
The first Royal motor journey, depicted here in a contemporary painting of the party, possibly at Warwick Castle, in June 1898. HRH the Prince of Wales (centre) is seen accompanied by the Duke of Marlborough, Lady Randolf Churchill and the Countess of Warwick. J.S. Critchley, the driver on this occasion, standing by the side of the 4hp, 2 cylinder Daimler car.

Left:
Gottlieb Daimler in the car that he constructed in 1886. His son Adolf is at the tiller.

Right:
F.R. Simms pictured in England in his Cannstatt built Daimler car.

Facing page:
The workforce and products of the Cannstatt factory 1898. This picture illustrates the wide range of products being manufactured at Cannstatt at the time.

Simms had been born into a highly respected Warwickshire family on the 12th of August 1863, a family which had been deeply involved in many different forms of mechanical engineering for several generations. Like his forebears, Frederick had a natural mechanical aptitude and served his apprenticeship at an early age in order to develop his skills and knowledge. His passenger cablecar system marked quite an achievement for one so young, but in addition, he was an excellent draughtsman, as well as being a recognised expert in patent law.

Gottlieb Daimler was born in the small town of Schorndorf near Stuttgart on the 17th of March 1834. At an early age, he was apprenticed to a local gunsmith, where he showed an exceptional talent for the trade. However, despite pleas from his employer for him to remain at Schorndorf, Daimler decided his future lay elsewhere, and departed for the City. Not long after, he was on the move again, this time to Gräfenstadt where he joined the Werkzeugmaschinenfabrik Company. He continued with that firm for around three years, before going on to study engineering back in Stuttgart at the Polytechnic College from 1857 until 1859. He then came to England, and for a time found work at Sir Joseph Whitworth's works in Coventry and Birmingham.

It was not until Daimler reached his 38th year, that he first became closely involved with the internal combustion engine. In 1863, Doktor Nikolaus August Otto supported by the wealthy Eugen Langen, had set up his workshop in Deutz near Cologne, for the sole purpose of developing this new type of engine. Otto offered Daimler the position of Technical Director, which he accepted in 1872. One of Daimler's first actions was to secure the services of Wilhelm Maybach, whom he had known whilst Maybach was the foreman at a factory in Reutlingen. Maybach was soon appointed Chief Designer for Otto & Langen, and from then onwards, the two became inseparable companions.

The partnership between Daimler, Otto and Maybach continued until 1881, when Daimler left for Russia, but he returned a year later, and decided the time had come for him to set up his own private workshops. These he established at Cannstatt for research and experimental work with his old friend, Maybach. He purchased a house with a garden large enough for a workshop to be built at 13 Taubenheimstrasse, and it was here that the two continued their work on engines right up until the time of the Bremen exhibition.

Daimler's faith and enthusiasm in his invention was great, but he was careful to avoid expressing his opinions on unsubstantiated facts. However, the more Simms and Daimler talked, the more excited Simms became about the engine. He learnt that one unit had successfully been fitted to a launch and tested on the Neckar, and how a year or so previously, Daimler had driven through the streets of Cannstatt in a locally built vehicle powered by a Daimler engine at the rear; a simple steering system replacing the traditional horse shafts. An early motorcycle had also been produced by the Daimler and Maybach team, and a fire engine had employed the power-unit for use as a pump.

Simms secures Daimler engine for Britain

The single cylinder four stroke engine was beautifully made, ingenious and in some respects, a revolutionary design. One of the outstanding features was the ignition system which Daimler himself had invented, that later went on to be known as hot tube ignition; the main advantages being that it was self-timing and had no moving parts or sliding valves. A surface carburettor was employed, linked to an automatic inlet valve, and a centrifugal governor (which acted upon the exhaust valve), was also fitted. Even the piston itself incorporated a small valve - a novel feature to aid the escape of waste gases. With a bore and stroke of some 70mm x 120mm, the governor allowed the engine to run at a constant 750rpm to 800rpm.

Simms became a regular visitor to Daimler's house in Germany (the Daimler Motoren Gesellschaft had been founded in November 1890 with a capital of 600,000DM) and eventually, an agreement was drawn up giving Simms control over all the patent rights of the Daimler engine for the whole of the United Kingdom and all of her colonies with the exception of Canada.

In June 1891, Frederick Simms and his partner,

Right:
The Daimler powered boat first used on the River Thames in 1891.

Bottom of page:
This advertisement clearly dates the Bremen Show as 1890. This has previously been thought by some historians to have taken place in 1888. It also gives an indication of the range of products built by the Daimler works in Germany at that time.

Alfred Hendriks, established the firm of Simms & Company - Consulting Engineers, with offices at the Billiter Buildings in London. Unfortunately, the partnership was short and unhappy. It lasted until the end of the following year, when Hendriks decided to leave the country hurriedly, landing Simms with a chaotic state of affairs to clear up as best he could.

The patent rights to the Daimler engine were of little use to Simms at this time, due to the Highways and Locomotives Act which had been passed in 1878. The Act would not allow any self-propelled vehicle to use the roads except under quite impossible conditions. It was obvious to Simms, that until the Act was withdrawn, or at least revised, there was little point in trying to exploit the engines worth. In 1891, there was still nothing to suggest its abolishment in this country and, even on the Continent, motor vehicles were not taken very seriously. A few firms were making experimental cars in France, and in Germany, only Daimler and Benz were giving serious thought to the matter. All Simms could do was carry on as a Consultant, and in the meantime do everything within his power to promote the Daimler engine for use in launches, for pumps and general stationary work, and to promote the cause of the horseless carriage whenever possible.

However, Simms' enthusiasm for the new Daimler engine was well illustrated when he wrote to Hendriks during February 1891, telling him that he had started up a department within the company for petrol engined motorboats and cars, and that he wanted to exhibit a motor car at the German Exhibition at Earls Court, London. He also wanted to demonstrate a motor boat on the Serpentine, *"to show the grand thing in England which I have secured. The motor is quite unequalled. It has no smell, is quite dangerless, 50% cheaper than all other motors and with very little consumption .... it has many advantages."*

The first Daimler engine ever to come to this country arrived on 16th May, 1891. On the same day, Simms brought over the first Daimler-engined motor boat, which was loaned to him by the Cannstatt firm. Simms had made arrangements with Hartwig & Vogel of Dresden to use the Daimler engine to drive one of their chocolate and cocoa making machines at the Earls Court exhibition, but the sponsors raised all sorts of objections and the idea was abandoned. So too, was the idea of showing a Daimler car, as it was still illegal to freely demonstrate the vehicle to interested parties on public roads. Yet more disappointment was experienced with the Daimler boat. Simms was told that he would not be allowed to give demonstrations on the Serpentine in Hyde Park, so he tried to arrange the use of the lake at the Welsh Harp in Hendon, and the large lake at the old Crystal Palace, but again without success. As

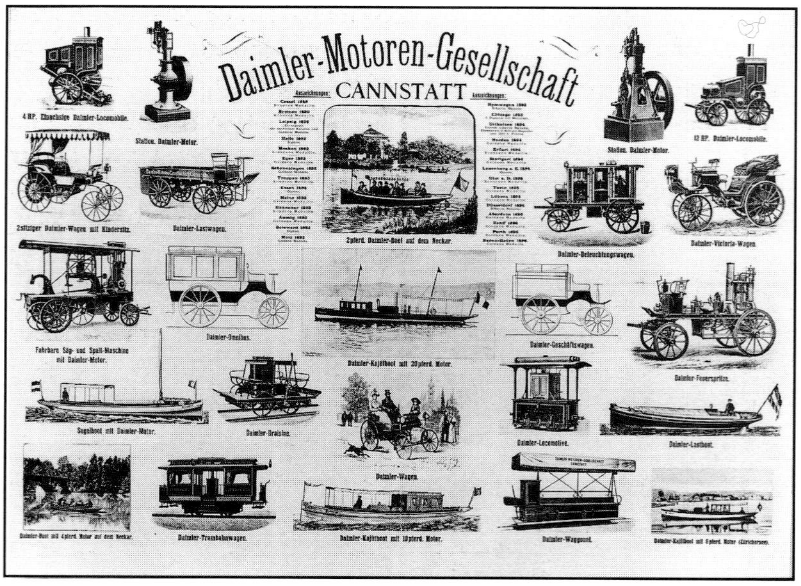

Left:
Daimler's belt driven car for the years 1892 to 1897. Powered by a water cooled vertical twin of 762cc, it gave 2.5hp at 700rpm and could achieve a speed of 12mph.

soon as it was known that some new type of benzene was going to be used, those in authority sensed the risk of fires and explosions. However, Simms did at least get permission to use the Thames, and many runs from Charing Cross and Westminster were carried out over the following few weeks.

The circle of motoring enthusiasts at such an early date was, by necessity, a small one, and the members usually known to each other. They met to discuss the progress being made on the Continent and the latest developments in this country, convinced it was only a matter of time before this new form of transport would be welcomed on the roads. There was a band of powerful people amongst them who tried their best to get the motor car recognised; one of these was the Hon. Evelyn Ellis, a member of the de Walden family, and later one of the first Directors of the Daimler Motor Company. Another was Henry Hewetson - a London tea merchant, and the first man to ever bring a petrol-driven motor vehicle into Great Britain.

The Daimler Motor Syndicate Limited

From 1891, when Simms established his business, until 1893, considerable progress was made with the Daimler engine and its many applications. Simms made repeated visits to the works at Cannstatt, and was eventually appointed a Director of the Daimler Motoren Gesellschaft. A number of motor boats were sold, and the engine was supplied for several other purposes. So, early in 1893, Simms decided to form a small private company with a small group of his friends for the sole purpose of promoting the Daimler products, thus paving the way for expansion into motor cars as soon as the law allowed it to become viable.

Simms got in touch with his cousin, Robert Gray, and a friend called Theo Vasmer, and they, together with one or two others, formed the Daimler Motor Syndicate Limited to take over the business of Simms & Co. All this happened during May 1893, with the capital of the new company being put at £6,000. The syndicate acquired all the goodwill of the Daimler business built up by Simms, including all the contracts, agreements, and sole selling rights for Britain of the patents held by the Daimler Motoren Gesellschaft, for which Simms was given £3,000 in shares, and just over £560 in cash. It was decided at the new company's first meeting, that they should rent a vacant arch at Putney Bridge Railway Station for the princely sum of £25 per annum, and use the accommodation as a workshop for the fitting of engines into river boats.

J. van Toll was appointed Chief Engineer, with two assistants. A better office in the Billiter Buildings was taken later on in the year and things seemed to go fairly smoothly for some time. At the end of 1894, Simms announced to the Board that he was resigning from his Managing Directorship, whereupon he was given the title of Honorary Director, with his place being taken by an Edward Atkinson. However, Simms returned to the Board during April 1895 as the Managing Director. By this time, there was some prospect of the law being altered in favour of motor cars taking to the road, and so the company took several steps to be ready for that occasion. Mid-way through 1895, it was noted

Above: *Harry Lawson on his Daimler engined Panhard, seen here at the end of the Emancipation Run. In fact it was Otto Mayer (seen in the picture standing behind the front right wheel) who drove the car for the entire run.*

in the Board meeting minutes:

"It was resolved to authorise Mr Simms to take all necessary steps in the face of good prospects, etc., for preparing everything towards the formation of the DAIMLER MOTOR COMPANY LIMITED, with a capital of £50 to £100,000, as may be decided later on, for the purpose of acquiring the British Daimler patents and the manufacturing of the said motors in this country".

The first car - a French Daimler engined 3.5hp Panhard & Levassor - was brought into this country on 3rd July 1895 by Ellis and Simms and, two days later, given an initial (and illegal) 56 mile trial run from Micheldever in Hampshire to Eliss's house in Datchet, Berkshire. Reports of this are said to have resulted in over eighty enquiries; some were orders for carriages, and others were offers to make the motors and have licences granted. Whichever way one looks at it, the decision to import the car and the resulting publicity could only be regarded as a complete success.

Lawson enters the scene

Every effort was made to hasten the formation of the proposed new company. The purchase money was fixed at £40,000, and the Hon. Evelyn Ellis was appointed a Director; so too was Gottlieb Daimler, who would hold the titles of Honorary Director and Consulting Engineer. The prospectus was well in hand and all of the officials had been chosen and duly elected, but then a move from a totally unforeseen quarter brought the plans to a halt.

Simms received a visit from Steadman & Van Praagh - a firm of solicitors hired to represent a private syndicate consisting of Ernest Hooley, Martin Rucker, and a certain gentleman called Harry Lawson. They wanted to purchase the British Daimler motor patent rights, and for this, they offered Simms £25,000 with a cheque for £15,000 on the spot. Such an offer was not to be taken lightly, and Simms put the matter before his Directors. After negotiations, it was agreed that £35,000 was to be the purchase price, £15,000 of this was to be deposited on account, and be forfeited should the balance not be paid on time.

It did not take long to complete the deal and Messrs Hooley, Rucker and Lawson promptly formed the British Motor Syndicate Limited to which all the Daimler patents were duly transferred. The shareholders in the original company saw a return of their capital plus 200% in dividends, and presented Simms with a magnificent silver cup. Henceforth, the newly formed British Motor Syndicate was free to do what it liked with the Daimler patents.

Of the three characters at the head of this new concern, it was Harry John Lawson who was by far the most interesting. He had drive and ambition, and was undoubtedly one of the most remarkable men ever to dabble in the motor industry. Born in 1852, he had already become exceedingly rich through the cycle and related industries, floating one company after another.

However, once the horseless carriage made its first appearance on the Continent, he saw that there was a vast fortune to be made for any financier with ample means, sound judgement and plenty of showmanship. His scheme was to buy up every promising patent before the British motor industry got started, in the hope that one day he would own some sort of master patent which in years to come would prove to be a gold mine.

The Daimler Motor Company Limited

Lawson was essentially a man of action; his plans for exploiting the Daimler patents, should he succeed in acquiring them, were in an advanced stage before such patents actually became his property, so no time was lost in floating yet another new concern called the Daimler Motor Company Limited, registered on 14th January 1896.

Seldom in the history of finance can a company have been promoted under such disadvantageous conditions and with a less hopeful outlook. The proper use of motor vehicles on the highways was still illegal, but Lawson was undeterred by this, and the following statement was made in the new company's official prospectus issued on the 17th of February 1896:

"Horseless Carriages worked by the Daimler motor have been and are running on the public roads in certain parts of England, but as in certain localities, the Authorities have expressed opinions restricting their use, it should be stated that special legislation for licensing them in every part of the United Kingdom is now pending, and a small tax will probably be imposed by Parliament, similar to that now paid for private carriages."

Apart from the odd experimental car from Wolseley (built by their young Works Manager, Herbert Austin) and Dr Fred Lanchester, there was hardly a man in this country with any sort of experience in either motor car design or construction, but not only were some of the original Directors of the Daimler Motor Company inexperienced, some of them didn't even know how the motor car worked. However, the following Directors were named in the prospectus quoted from earlier; Gottlieb Daimler, the Hon. Evelyn Ellis, William Wright JP, James Mace, J.J.H. Sturmey, and H.E. Sherwin Holt MA.

This was the Board nominated by the British Motor Syndicate as promoters of the Daimler Motor Company, but it was set out in the contract that they had the right to elect two other directors. Thus, Lawson later appointed himself as Chairman, and the other seat was given to J.S. Bradshaw. From an engineering point of view, it would have been difficult to find a Board less well equipped for the complex duties which lay ahead. Lawson was a financier who wished to be known as an engineer, having made claims to inventing the safety cycle. This may or may not be true, but one thing is certain, his chief engineering feat prior to the coming of the motor car, had been the engineering of his own fortune.

Unfortunately for the new business, Daimler himself was a Director in name only; he never attended a meeting and took no part in the firm's affairs. After his retirement from the Board the title of Honorary President was bestowed upon him, although in reality this had very little meaning. Ellis was a keen amateur, but had little in the way of practical knowledge in any form of engineering - Wright and Mace knew even less. This left Sturmey. Born in Somerset in 1857, Henry Sturmey was originally a schoolmaster who later turned his attentions to cycling when it was first introduced, and founded the newspaper known simply as *The Cyclist*. He became the first Editor of the well-respected journal *The Autocar,* and although he had a certain flair for engineering, his knowledge of motor cars was, at that time, somewhat limited to say the least.

The registered offices of the Company were shown as the Billiter Buildings, London, the same address as the old Daimler Motor Syndicate, with Charles Osborn (who had worked for Lawson before), appearing on the prospectus as its first Company Secretary.

As well as the obstacles brought about by inexperience, there was a growing reluctance from the public to invest money in anything that Lawson was involved in. Both he and his various companies were finding themselves increasingly under attack from the financial press. Articles were beginning to appear that outlined the amount of concerns that Lawson had floated against the number which still existed. Indeed, the *Stock Exchange Gazette,* writing in the early part of 1896, went so far as to say *"the fact that Mr H.J. Lawson is the controlling spirit is a very bad omen for the company, and augers a speedy acquaintance with the Bankruptcy Court rather than the success the patent deserves."* In addition, there was still a legal embargo on the full use of motor cars, resulting in conditions which seemed to spell little hope for the flotation of the Daimler Motor Company, let alone its future thereafter.

When the time at last came for the flotation, the heavens must have been smiling on Lawson, for there was in fact an over subscription of some £10,000. The capital of £100,000 was divided up into 10,000 shares of £10 each; £60,000 was put by as working capital, while £40,000 of it was paid to the British Motor Syndicate Ltd for a licence to manufacture under the Daimler patents. These terms were laid down under an agreement dated 30 January 1896, between the two contracting parties, stipulating within the document that the consideration should be either 4,000 shares, or cash in lieu thereof. Lawson undoubtedly decided that the in lieu thereof option was undeniably the more attractive, and once again he had the last laugh.

It seems that Lawson could do no wrong. As Montague Grahame-White once said of him, *"Short in stature, high in self esteem, and 'tall' in finance, he was clever enough to extract from the investing public over two million pounds for his company promotions within a period of ten months."*

Critchley joins the Company

With such a Board a Directors, it was essential if the Daimler business was to be any sort of a success to appoint an experienced engineer. It was Henry Sturmey who introduced J.S. Critchley, and after a short interview with F.R. Simms (who held the position of Consulting

Above: *An advertisement for the Motor Mills, Coventry, showing a picture of the works.*

Below: *An advertisement dated 1896 for the Great Horseless Carriage Company, also based at the Motor Mills.*

Engineer), James Critchley was given a three year contract worth £400 per annum, and the important role of Works Manager with Otto Mayer being brought over from Cannstatt to assist.

Born on the outskirts of Birmingham in 1872, Critchley attended King Edward's Grammar School in that city and excelled as an athlete. Quite where he gained his engineering skills is unknown, but to be offered a job of this importance at the age of just twentyfour obviously showed a great deal of faith in the young mans talent. In the years ahead, Critchley would have a massive bearing on the fortunes of the Daimler Company.

Once all of the legal formalities had been completed, some of the Board members went over to the Continent where they could see for themselves how the work at some of the larger French and German factories was conducted, and what was being produced. The Panhard & Levassor works were visited (Panhard held the Daimler engine patents for France incidentally), along with those of De Dion-Bouton. While they were in Paris, an order was placed for two complete Panhard cars to be delivered each month, with the order later increasing to six per month. After calling in on the Peugeot factory, the whole party moved on to Cannstatt to see Gottlieb Daimler. They were shown around the old Daimler Motoren Gesellschaft factory (which was later destroyed by fire in 1903), and the British contingent came away with a strong feeling that a suitable factory should be found back at home in order to start car manufacture as soon as possible.

The Move to Coventry

The search for premises began in earnest once the party returned to England. Possible sites were looked at in both Cheltenham and Birmingham without any success but then, largely by chance, they heard of a disused cotton factory which had previously been used by the Coventry Cotton Spinning & Weaving Company. During 1891, there had been a fire which had burned out the top three floors, but repairs by the insurance company had resulted in virtually new premises lying idle. The site consisted of thirteen acres of leasehold land with a double factory on it, but due to its size and an ailing textiles industry, the property had proved to be something of a white elephant; after three years on the market, not a single customer had come forward. This useful discovery was made in the April of 1896 - not wanting to build, the Directors saw their chance and immediately snapped it up.

It just so happened that at the time, Lawson was on the verge of floating yet another motor car company with the ambitious title of the Great Horseless Carriage Company, and this new concern would also be in need of factory accommodation. The second building on the estate was precisely what was needed for Lawson's requirements, and the upshot was that the second building was sold to the Great Horseless Carriage Company for rather more than the total sum that Daimler had paid for the entire site. This was quite a useful beginning, and an unexpected windfall which provided extra funds for equipping the works.

An idea of the magnitude of Lawson's financial dealings can be gauged from the fact that the Great Horseless Carriage Company was offered to the public for subscription with a capital of £750,000. According to the prospectus (which mentioned Gottlieb Daimler, the Comte de Dion and E.J. Pennington as Directors), two thirds of the total was in respect of a contract entered into for licences to use certain patent rights belonging to the British Motor Syndicate. The new concern also had a contract with the Daimler Motor Company enabling it to purchase Daimler engines at a discount some 10% better than any other customer. It should be remembered of course, that Lawson had been the instigator of all three firms, so stood to gain from every angle!

S.F. Edge said of Lawson at the time, *"He paid away £10,000 as the ordinary man would a £5 note; like many men whose previous ventures have met with a marked success, it did not occur to him that he could make one or two false moves which would mean his fortune slipping between his fingers."*

When the Daimler Motor Company first took over the factory, most of the shops were still filled with the old cotton-spinning machinery, which was gradually broken up and cleared away to make room for the new plant on order. The Daimler premises merely had four walls and some overhead shafting which was left in place by the new occupants. Most of the new machinery was ordered almost straight away, but quite a large proportion of that ordered in May did not arrive until December 1896, whilst other pieces were still arriving as late as the following March and April. In order to overcome this, the Company ordered 100 complete chassis from Panhard & Levassor in June 1896, enabling the Company to start "production".

The first statutory meeting of the Company

The Company's first statutory meeting was held on the 19th May 1896 at the Institute of Chartered Accountants, naturally under the presidency of Lawson.

Left:
A Peugeot 4hp, 5 seater 'Omnibus'. This vehicle is of the type sold by Daimler in England and exhibited on their stand at the first London Motor Show.

Facing page: *Taken shortly after the 'Emancipation Run' this picture shows some of the people involved in the Daimler motor Company in the early days.*
Standing l to r: Otto Mayer, J.S. Critchley, An unknown, Walter Arnold, Thomas Robinson, Charles Turrell, Turrell's father, H.O. Duncan, and A.J. Drake.
Seated l to r: Charles Osborne, An unknown, H.J. Lawson, Lawson's eldest son, Tom, J.H. Mace and Charles Harrington-Moore.

It was held merely to conform with the requirements of the law, with little information being given beyond the details of the journey made to the Continent. However, it was stated that further works had been purchased in order to carry out motor launch work. These were situated on Eel Pie Island near Richmond, and were considered to be ideal for the purpose.

The whole statement was essentially an optimistic one, with an announcement that the Company already had *"enormous orders"*, and that whilst the Coventry factory was still not ready, cars and engines would be imported to fill contracts. In short, it was merely a question of waiting a little while until the payment of healthy dividends.

In the meantime though, a black cloud had been gathering in the background. The optimism of Lawson and his fellow Directors was not shared by some sections of the Press, who quite rightfully pointed out that the value of Daimler shares had fallen already. However, this was hardly surprising, as a legal embargo still existed on the Company's main product, and this could simply have been a case of Lawson bashing. Notwithstanding, a certain amount of anxiety spread through the shareholders, not helped by an article in *The Economist*, who noted that *"many* [of Lawson's company's] *have long since been buried and almost forgotten, except by surviving sufferers."*

An Extraordinary General Meeting was called during March 1897, and Henry Sturmey was invited to speak to try and ease troubled minds. He pointed out that a very large sum of money had been invested in the factory and that new and expensive machinery had been needed. It was also noted that the premises in Holborn Viaduct had been vacated in favour of much larger showrooms on the Shaftesbury Avenue. For the time being at least, Sturmey had managed to quell the unrest, but results would be needed soon.

London's first Motor Show

Shortly after the birth of the famous motoring journal, The *Autocar,* there was intense activity to campaign against the *"absurd restriction"* imposed on motor cars, and to get Parliament to introduce legislation which would allow for their proper use on the British highways. After all, nowhere else in Europe was restricted to this extent, and France even allowed motor races on public roads. A petition was launched by the magazine in February 1896, and an exhibition arranged at the Imperial Institute in London so that people could have an opportunity to see a motor vehicle first hand on one of the dates set between the 9th of May and the 8th of August 1896. This was the first motor exhibition to be held in London, so it was hoped that society would flock to see the event, particularly when they learnt that demonstrations were being given in the private grounds that adjoined the Institute. Invitations were given to both Houses of Parliament, as well as to a large number of influential people in the hope of gaining their support.

At the show, the Daimler Motor Company occupied three stands and exhibited a *"Daimler oil motor omnibus constructed by Messrs Peugeot Frères of Paris"*, a *"Vis-a-vis Daimler motor carriage"* also built by Peugeot, and a pair of Daimler Motor Victorias - both of which being fitted with a new 4hp engine. Although this information is not exactly correct (Peugeot did not have a Paris factory until 1904), it does at least give some idea of what Daimler had to offer at the time. In another room, the Company displayed Gottlieb Daimler's experimental quadricycle which he had built in 1886, as well as one of his early *"motor horses"*, a 2hp Daimler tramcar and tender, a Panhard & Levassor chassis frame, and a *"Motor omnibus to seat four persons and driver"* from the same French concern.

The first Royal motorist

When His Royal Highness the Prince of Wales, later King Edward VII, expressed his desire to see and try one of the machines on show at the Institute, he was given a private viewing long before the exhibition opened to the public, and driven by the Hon. Evelyn Ellis through the gallery and out into the grounds on F.R. Simms's belt driven Cannstatt Daimler.

This historic event took place on February 14th 1896, and signified a turning point in the future of the British motor car. The Prince was delighted with this new experience, and despite being a keen horse lover, had no qualms when it came to showing support for the automobile. Indeed, he was later acknowledged as saying, *"I shall make the motor car a necessity for every English gentleman."* This support had perhaps come a little too late, as Britain was already lagging behind its Continental competitors by a number of years. As Thomas Edison, the extraordinary inventor, once said: *"The motor car ought to have been British. You first invented it in the 1830s. You have roads second only to those in France. You have hundreds of thousands of skilled mechanics in your midst, but you have lost your trade by the same kind of stupid legislation and prejudice that have put you back in many departments in the electrical field."* As a good friend of Henry Ford (they actually bought houses next to each other in Florida), he could have been sharing the ironic joke with Ford himself.

The problem stemmed from the short sightedness of the British Government. The 1878 Act had been drafted by the Local Government Board despite admitting to the fact that they *"had no experience of these light locomotives."* The Locomotives on Highways Act was passed through amid laughter following Henry Chaplin's suggestion that one day *"these motor cars might become a rival to light railways."*

The biggest worry seemed to be that the automobile would *"ruin the horse breeding industry"*, but one only had to look across the Channel to see the vast employment created by this new field of activity - obviously, most backbenchers had never heard of Panhard & Levassor, Peugeot and Benz. In France alone during 1896, the car industry had sales of 1.3 million francs, but in England it was to be some 60 years later that someone pointed out that *"in this mechanical age, to restrict a country's motor-mindedness is to condemn it to technical backwardness."*

However, the show was by all accounts a great success and the passage of a new Bill through Parliament was assured, especially now that Royalty was showing active support. Lawson and the various companies in which he had an interest, under the banner of the so-called Motor Car Club (founded by Lawson for *"the Protection, Encouragement and Development of the Motor Car Industry"*), began to make arrangements to celebrate the coming of the Locomotives on Highways Act on the 14th of November 1896.

A little known fact is that Lawson actually tried to schedule a run from London for the 24th October, *"to give the officials of the Local Government Board*

Right:
A Daimler parcels van dating from the latter part of 1897.

Facing Page:
J.S. Critchley at the tiller of the first Coventry built Daimler car, pictured during the first few months of 1897.

engaged in framing the rules governing motor traffic and representatives of public bodies a practical demonstration of the capabilities and characteristics of the new vehicle." Despite participants being told that *"motor cars are on trial in England, and that any rashness or carelessness might imperil the industry in this country"*, the Surrey Police refused to give permission for the event until the new Act came into force.

The sad thing is, that although Lawson wanted to promote the industry for his own (and largely financial) reasons, there was a lot of sense in having an organised demonstration to allow the LGB to put down some realistic guidelines. As it happened of course, the 12 mph limit was out of date almost at once, but it was a great improvement on the existing law.

The Emancipation Run

"Today, November 14th, 1896, is a red-letter day, not only in the history of automobilism, but in that of England itself, for it marks the throwing open of the highways and byways of our beautiful country to those who elect to travel thereupon in carriages propelled by motors, instead of horse-drawn vehicles or upon bicycles...." - The Autocar.

It was decided that there would be a procession from London to Brighton of every motor vehicle that could possibly be rounded up in time - several cars were brought over from the Continent just to take part, H.O. Duncan, Lawson's Paris Agent, offering incentives to anyone who owned a vehicle. Although the recently formed Daimler Motor Company could not take part with any cars of their own manufacture, it is interesting to note that many of the cars used were equipped with Daimler engines made under licence, and that at least the Daimler staff were very much in evidence. J.S. Critchley, the Works Manager, drove a Daimler parcels van all the way down to the capital from Coventry, and then on the following day completed the run to Brighton in the same vehicle, although it must be said that his journey was not without its troubles. Otto Mayer was chauffeuring Lawson and his wife in the Daimler engined 6hp Panhard & Levassor that had won the Paris-Bordeaux-Paris race in June 1895. Also on this most famous of London to Brighton runs was a car shared by Simms and Daimler. Sadly, Gottlieb Daimler never lived long enough to see the motor industry fully develop - he died of heart trouble on the 6th of March 1900, at the age of 66.

Léon Bollée was the first to make Brighton incidentally, getting there a long way ahead of the other entrants; his brother, Amédée, was second, on a similar Bollée machine.

The throwing open of the highways to motor cars was, of course, a great boost to the fledgling British automobile industry. From now on, motor vehicles would be allowed to use the roads without fear of breaking the law, unless exceeding the new speed limit set at 12mph or 14mph, depending on the locality.

The Company's first catalogue was issued in 1896, with the registered address shown as 95 Billiter Buildings, Leadenhall Street, London. It contained little more than illustrations of early Panhard and Peugeot cars which were still being imported, and there was nothing to suggest that the manufacture of cars in this country was even being contemplated. However, only a short while later, another brochure was released showing what the firm were hoping to produce in the factory at Coventry, by now known locally as the Motor Mills.

The first British Daimlers

Although Panhard-Daimlers had been sold through the Company as early as November 1896, a letter in the author's possession written by J.S. Critchley in April

1944, suggests that the first real Coventry built Daimler was completed in the early part of 1897. It had evidently been finished a little late, as according to A.H. Deaville-Altree, it should have been ready in time to compete on the 1896 Paris-Marseilles Race. However, another new catalogue was produced not long after the completion of this historic machine, and this was the first to show actual Daimler products.

Every vehicle described within the pages employed the same type of twin cylinder 4hp engine with tube ignition and chain drive. Each model was given a different name according to the type of body fitted, such as the Rougemont, the Siamese (built by Arthur Mulliner), the Universal, the Grafton, Wyley and the Jaunting car. Also in the catalogue were several letters apparently written by Daimler owners, though their views can hardly be counted as impartial as two of the signatures belonged to Directors of the Company!

The name of the late Major General Montgomery of Winchester will attain immortality in the history of the British motor industry, for it was he who was the first person unconnected with the trade, to place an order for a British built Daimler car. Indeed, there is good reason to believe that his was the first order from a private customer for any petrol engined car built in this country, for it was still only the August of 1897. Montgomery's particular car, which according to the *Hampshire Chronicle "was simply delightful, the swift, gliding, noiseless motion along level roads being particularly exhilarating"*, was originally built for Henry Sturmey, who intended to use it on a John O'Groats to Land's End journey, but as the Company was so far behind on its promised delivery dates, he gave up his mount in favour of the Major General and postponed his trip until the autumn when another vehicle became available.

Orders were coming in from a number of other concerns; Messrs Stirling's Carriage Works of Scotland were strong supporters of the Daimler engine, the Midland Motor Carriage Syndicate ordered an undercarriage for their steam vehicle, and both Burgess & Co. and the London Motor Van & Wagon Co. were supplied with Daimler frames and engines. Slightly later, 4hp chassis were also supplied to Charles Crowden the founder of the Leamington based Crowden marque, who built experimental cars and fire engines from 1898 to 1901.

However, Daimler were still buying in from other manufacturers themselves, even at this stage. Ten Bollée motors were received from the Great Horseless Carriage Company in April, and it seems that the GHCC were trying to sell complete Bollée machines to them as well, Lawson having bought the rights and drawings from Léon Bollée himself. The offer was refused on the basis of there being no guarantee as to their efficiency and reliability.

New Beeston tricycles on the other hand, were sold through Daimler at a retail price of £75. By mid-1897, however, no more were being ordered, and the Company was desperately trying to dispose of their remaining stocks, giving 15% discount to the trade and

Above:
July 1897 and a picture of the Daimler car that reached the summit of the Malvern Hills.

Below:
The twin cylinder 4hp car that Henry Sturmey used on his 'end to end' run. This car was capable of 13 mph in favourable conditions!

5% to private buyers.

In the meantime, Altree was trying to resolve the non-payment of the British Motor Syndicate account with Harry Lawson, something which dragged on for some time. In December 1897, Lawson even had the cheek to offer Daimler a £100 4% Debenture in this latest venture of his - perhaps fortunately, the Company declined, even though they did hold a number of shares in the BMSL already.

The fact that Daimler and the GHCC were indeed two separate concerns has been well established, but they did work closely together. A Board Minute of July 1897 stated, *"It was agreed to supply them* [the GHCC] *with one new set of drawings of the Panhard frame and gearing without charge."*

It seems odd then, that Daimler in Germany should be refused permission to see working drawings of the 4hp engine as made in Coventry, especially as the engine as it stood, apart from a very slight modification on the crankshaft, was identical to the German type. In July 1897 though, *"Drawings of German and Panhard types of Daimler engines were submitted by Mr Drake, and the Works Manager was instructed to put the manufacture of one of the latter type motors in hand at once."*

The climbing of Malvern Hills

During July 1897, the Hon. Evelyn Ellis and James Critchley performed the first of two remarkable feats in the early Coventry built Daimlers. Ellis's house was at the foot of Malvern Beacon, and he and Critchley set

themselves the task of driving a car to the top of the famous hill. The car was geared down specially for the occasion, as it would be carrying three people and would have to contend with gradients greater than 1 in 5 in several places. Ellis, his young daughter and Critchley, set off from Malvern Link and drove for nearly four miles before finally reaching the summit.

This was a truly exceptional performance for such an early car, when water cooling was still very primitive, and overheating was a common problem. The climb caused a mild sensation and was very good for publicity, prompting the Hon. C.S. Rolls to buy the No. 6 Carriage (one of the original stock) for "*not less than £300*".

Sturmey's End-to-End Run

Although Lawson was the official Chairman of the Daimler Motor Company, it was Henry Sturmey who usually presided at Board meetings. Born in 1857 and with a teaching background, he was certainly a valuable asset, for he was able to conduct some excellent propaganda work in the columns of *The Autocar*, and lost no opportunity in doing so, not only for Daimler, but for any concern in which Lawson had an interest. This undoubtedly had something to do with the fact that Sturmey had a large amount of shares in numerous Lawson concerns, including £14,000 tied up in the Great Horseless Carriage Company alone. Perhaps his greatest service to the Daimler Motor Company though, was the outstanding publicity he gave the firm in October 1897.

Sturmey's machine had a twin cylinder engine of 4hp rating with tube ignition, solid tyres, tiller steering, and a body by Mulliners of Northampton built to Sturmey's own design (Mulliners built most of the Daimler bodies at this stage, the cars being sent to Northampton by rail. But they were far from happy with their service, even going so far as to look at Thrupp & Maberly in London, as well as Windovers and Offord & Sons). Its range of speed was between three and thirteen miles an hour, and with the first AGM of the Company fast approaching, Sturmey was quite anxious to report something outstanding in the way of the vehicle's overall performance.

Accompanied by a mechanic from the Daimler works called Ashley (in all of the reports featured in the *Autocar*, there was never any reference to Ashley's first name), Sturmey took his car by train to Wick, then drove it to John O'Groats ready to start his long journey on Saturday 2nd October 1897. It was the first time that an automobile had been seen in many of the villages and towns en route, and as Sturmey knew this would evoke many questions each time he stopped, he carried with him a supply of cards to hand out, giving the particulars of his vehicle.

Not surprisingly, the chief difficulty in organising this 900 odd mile long trip was planning for the supply of petrol, as it was sold hardly anywhere at this time. The oil refiners Carless, Capel & Leonard were duly instructed to send supplies to various places through which Sturmey and his companion intended to pass, and as a result, there was remarkably little trouble experienced throughout the whole of the journey. One of the most exciting moments occurred whilst trying to negotiate a large hill near Ambleside in the Lake District. This is how Sturmey himself described the event:

"*Ashley was doubtful about negotiating it, but as we had to get down somehow, I thought I might just as well try this road as any other, so I brought the car to a standstill on the edge and then with both brakes half on, allowed her just to drop over, bringing my cycling experience into play and keeping her in hand from the start. We descended very carefully for a hundred yards, and then Ashley asked whether I had both brakes hard on, to which I replied: 'No - only half on and I can stop at any minute.' To prove the correction of the assertion, I did so and again proceeded, much reassured by the incident. Steadily and slowly we crawled down the hill until seven-eighth's of the journey had been completed, and just as we turned the last zigzag and there remained a couple of hundred yards of perfectly straight road in front of us, I suddenly felt the car moving on. Both brakes were promptly clapped on full, but in an instant we were bounding away at forty to the hour, and I grasped the fact that the car had run away! I took the situation in at a glance and seeing that the foot of the hill was followed by a stiff ascent, I felt that all I had to do was to keep a straight course, so jamming both brakes on with all my force (although this appeared to have as much effect as water on a duck's back), I set my teeth and gripping the steering handle, steered a dead straight course in as near a bee line as possible with the result anticipated, and the car pulled up on the opposite slope.*"

This makes amusing reading when taken in conjunction with Sturmey's remarks made at the AGM held only a week or so later on. "*On the point of running away,*" he said, "*I am open to assert that under all ordinary conditions it is absolutely impossible for a Daimler carriage to run away. We have no less than eight different means of stopping our motor or our car, and it is not likely that they will all fail at the critical moment.*"

On the 19th of October at 4.45pm, Sturmey and Ashley finally arrived at Land's End, having covered a total of 929 miles in just over 93 hours running time, or an average for the trip of approximately 10mph. This journey became the subject of eager discussion, and the fact that no serious trouble was experienced along the way, shows the high quality of workmanship put into the Daimler cars, even in those very early days.

For his trouble, the Company paid Sturmey his railway fare to Wick, and £40 expenses.

During 1897, there had been a great deal of activity at the Daimler factory, with several changes in personnel and some new appointments. A.H.D. Altree became the General Manager, and a certain E.M.C. Instone was awarded Altree's old position of Company Secretary. Instone's long career with Daimler began with this appointment and, apart from a couple of years absence, he was connected with the Company right up until his death in 1932, first in an administrative capacity, and later forming Stratton-Instone with another Daimler official, Undecimus Stratton. He was also well known for his racing exploits, and was a regular winner with Daimlers in various forms of competition during the Edwardian era.

It was also in this year that F.R. Simms resigned *"with regret"* as the Company's Consulting Engineer to concentrate on his own business again, although he did still do some work for Daimler, and used one of their power units for the *"Motor War Car"* that he designed for Vickers, Son & Maxim in 1902. Simms, often referred to as the father of the British motor industry having founded both the RAC and the SMMT, retired from the trade in 1935. He died in 1944 at the age of 81, leaving two daughters.

The first AGM

The Company's first year of operations had not been a great success financially. At least one unforeseen misfortune had been encountered, involving the production of motor launches fitted with Daimler engines. The project was found to be quite unworkable and it was actually cheaper for the Company to purchase the launches rather than make them themselves. This particular venture had proved to be something of a nightmare, so it was decided to sell the Eel Pie Island works, and to shoulder the losses sustained. In May 1897, the Island was put up for sale by auction with a reserve price of £3,500. In the meantime, the Company had ordered a prestigious 56 ft launch from Messrs Paterson at a cost of £587, and in June a 31ft Daimler 4hp engined launch had been sent from Twickenham to Spithead to be present at the Naval Review. The production of Van Rennes diesel launch engines was also briefly looked into, so it was obvious that the Company still considered this to be an important part of the business.

The fortunate acquisition of the factory at Coventry was balanced during the first years operations, by the strictly limited number of vehicles produced by the time the first AGM took place. Some idea of what mass production meant in those pioneering days can be gauged from Sturmey's remark that it was impossible

Below: *An MMC dog cart from around 1898. This is virtually indentical to the Daimler models built only yards away in another building.*

Above:
Mr Crowden, a local manufacturer about to set out on a journey from Coventry to Birmingham in his twin cylinder Daimler in 1898.

to produce a complete motor car, however well equipped and managed the factory might be, in less than two to three months. Even in 1900, Daimler could only manage to produce 150 cars for the year, despite aiming for eight per week.

This first AGM was held at the Holborn Restaurant on the 14th of October 1897, while Sturmey was still engaged on his John O'Groats to Land's End journey. As it happened, he left the car somewhere in the Midlands in order to attend the meeting. Sturmey took the chair as usual, and after reading a letter from Lawson announcing (probably to the relief of many) that he was resigning from both his position as Chairman and from the Board, he went on to deliver an optimistic speech. It was explained that during one period of the year under review, the Company had run dangerously short of capital due to the costs involved in fitting out a new coachbuilding facility, but that the Hon. Evelyn Ellis had promised to help financially should things become really uncomfortable.

J.S. Bradshaw, a Director and shareholder, reported on a 130 mile trip which he had made from Coventry to Liverpool, complete with seven passengers and their luggage. After stating that he had averaged 12mph, he summed up by saying: *"I think this speaks volumes for our Daimler carriages, as it was a severe test for a motor of only 4hp. In all, this particular carriage has run about one thousand miles, and to my mind the Daimler motor car is almost perfect."*

Those at the meeting seemed quite happy that although no dividends were paid out this year, at least the future held some good prospects. The factory had been acquired at no cost at all, and was now running as efficiently as could be hoped for at this stage. There appeared to be every justification in Henry Sturmey's optimism, although in the privacy of the Board Room, Sherwin Holt had passed the resolution; *"That the output of the Works is totally inadequate for their size, and the work itself is defective. That in the opinion of the Board, entire re-organisation is advisable and that steps be immediately taken to obtain an experienced and energetic Works Manager to supersede Mr Critchley either temporary or permanent as may be found advisable."* This was a fact recognised by Ellis, too, as he offered a £10,000 advance to the Company *"provided thoroughly satisfactory work was turned out."*

Although undeniably of quite a simple design, the British built Daimler never tried to deviate from a school of engineering thought that was being followed closely by all the leading manufacturers on the Continent. For such a young company, this was a wise decision which allowed it to develop from the experience of others. Limited means and the need for economy prevented the Company from indulging in costly experimental work that might well have proved a failure anyway, but in spite of this, no steps in car design were allowed to go unnoticed.

With four cylinder engines for heavy cars very much en vogue on the Continent, the Daimler Company followed this lead, and during the last months of 1897 produced their own similar power-unit of 8hp. The cars powered by this new engine were said to be capable of *"very high speed indeed"* to quote one authority, and the first to be produced was bought by a Professor Boverton Redwood. There is good reason to believe that this was the first four cylinder car ever to be built in this country. A number were also supplied to the GHCC, but things were soon to change in that direction.

The Great Horseless Carriage Company

Unfortunately, the Great Horseless Carriage Company was by this time already in serious financial trouble; still nothing was heard of E.J. Pennington, Lawson's other hope in the GHCC building, and storm clouds were gathering on the horizon for Daimler, too.

The GHCC shareholders had formed a Committee, sending out a strongly-worded circular in October 1897 - their £10 shares at this stage were only worth a few shillings. The BMSL and the Daimler Company were described as enemies, Daimler *"living freely upon us,"* which to a certain extent was right of course, due to the factory sale. The well-informed were not surprised, as this was a Lawson trademark; he had no interest in making a business work, just making fast money.

Shortly afterwards, in the early part of 1898, the Great Horseless Carriage Company Limited was re-formed and named the Motor Manufacturing Company Limited (or MMC), some of their products being almost identical to those of Daimler, not only in their appearance but in their chassis layout as well.

Another Royal occasion

A certain amount of confusion still exists today with regard to the association of the late King Edward VII - then the Prince of Wales, with British motoring history in its very early days. It has already been stated that the first time His Royal Highness tried a motor car was at the Imperial Institute in 1896. He also came into contact with a Daimler when the Duke of York used one at a shooting party sometime in September or October 1897.

This somewhat brief trial was followed by a proper demonstration in the grounds of Buckingham Palace on the 27th of November 1897. On this occasion, the cars consisted of Sturmey's Daimler that he had used on his epic end-to-end trip, another Coventry built Daimler (this one belonging to Arthur Mulliner of Northampton), and two motor tricycles; one a Beeston and the other a De Dion machine. The Prince tried both cars and was said to be pleased with their running.

The first time the Prince ever tried a car on the public roads, however, was in the June of 1898 during a visit to the Earl of Warwick at Warwick Castle. As soon as he expressed his wish to go for a drive, the Daimler Company were notified, and they quickly assembled a selection of cars. There were three standard 4hp twin cylinder models, Boverton Redwood's four cylinder car, and Sturmey's end-to-end Daimler again. The Company took no chances regarding the drivers for this auspicious occasion, so J.S. Critchley accompanied by A.H.D. Altree, gave various members of the party rides around the grounds.

As it happened, a visit had been planned for that day to Compton Verney, the seat of Lord Willoughby de Broke, and so instead of going there by carriage, it was decided they would go by motor car. With the four cylinder delivered to its owner in nearby Leamington Spa after a brief demonstration, Critchley drove the Prince, the Duke of Marlborough, the Countess of Warwick, and Lady Randolph Churchill on towards Compton Verney. The car driven by Critchley ran poorly due to a partly blocked silencer, and some

Above:
An 1898 Daimler currently on display at the Museum of British Road Transport in Coventry.

assistance was called for when a stiff rise had to be negotiated on the return journey. After making their way to Wellingbourne House, the residence of William Lowe, a well-known race horse breeder, sadly the weather broke, and it was decided to complete the journey in closed carriages rather than face the elements in open motor cars.

Critchley, writing in 1911, stated: *"I often think what a misfortune it would have been to motoring generally if some serious breakdown had occurred but, as matters turned out, the King's verdict was favourable to the new movement, and the automobile was destined to enter most completely into his life. Indeed, the influence of his tactful use of the motor carriage cannot be exaggerated."*

Other News

Panhard & Levassor obviously gave one or two orders to Daimler during the early part of 1898, perhaps either because their own production was over-stretched (they built 336 cars in 1898) or maybe cars ordered by English clients were produced by arrangement.

The Coventry Autocar Syndicate Ltd approached Daimler regarding to the supply of engines, and *"Mr Holt reported that he had received a letter from Mr Daimler with reference to the proposed importation of Cannstatt carriages by Mr Simms. Mr Holt was authorised to inform Mr Daimler that the Board were desirous of exchanging particulars of improvements direct with the Daimler Motoren Gesellschaft."* Sales to India were looked into via a Mr Digby, and the Directors and management of the Motor Manufacturing Company visited the Daimler Works in order to exchange information.

An Experimental Shop was established in April 1898 following Mr Holt's suggestion, after the failure of numerous components. The new department brought about improvements in the steering gear, and also prompted the Company to introduce the modern idea of sealing the cylinder head with a copper joint.

As well as F.R. Simms putting forward his electric ignition for testing, a new style of improved hot tube was also investigated, known as the *"Silent Lamp"*. Designed by a Mr Lyons, a number of flaws were found, and Critchley (still in charge despite Holt's letter) set about trying to improve it, although it was never adopted on a production car. After a number of months, it was eventually abandoned.

The 8hp machines (their engines were basically two coupled 4hp units, copying the Maybach and Levassor "Centaur" principle, ie. linking two "Phoenix" V-twins to give four cylinders) underwent a great deal of development. New gear ratio's were also tried, giving the car speeds of 3.25, 6.5, 9.5 and 13.5mph through the gears. So successful was the 8hp, that eventually a night shift was introduced to cope with demand, as if the hours were not long enough already; 7.30am - 12.30pm then 1.30pm - 6.00pm Monday to Friday, and 7.30am - 1.00pm on a Saturday.

Daimler 4hp engines were supplied at the rate of two a week to the MMC, and at this time, a decision was made to fix a plate to all of the cars supplied to the BMSL - *"Daimler Motor Company Limited, makers, London and Coventry, licensed by the British Motor Syndicate Limited."*

Lawson then approached Daimler regarding the supply of English cars to France via Panhard & Levassor. He was quoted a normal wholesale rate, but also told of the Paris firms previous communication on the samesubject. The Board's trust of Lawson was obvious, however, one-third payment on order, with the balance prior to delivery! In reply, Lawson continued to openly use Daimler patents without paying further royalties. The Company were not very happy, to say the least, and lodged a formal complaint - no money was forthcoming, however.

The name "Motor Mills" was officially dropped in favour of "The Daimler Works" in June 1898 and in the following month, George Iden, the Manager of the MMC, was approached regarding the latter supplying bodies to Daimler, for as yet, they did not possess a Body Shop.

The Motor Manufacturing Company later looked at buying the complete 8hp chassis from Daimler. Although they returned it after a detailed examination, J.H. Gretton, MMC's Chairman, ordered three 4hp engines a week, and Daimler bought many castings and their wheels from MMC, so it was a good working relationship. Constant financial bickering over invoices, MMC's complaints over *"the alleged insufficiency of the hp developed"* by Daimler engines, and the attempts by MMC to remove Daimler nameplates and stampings on their cars surprisingly did little to destroy it.

Towards the end of 1898, at least two engines were converted into diesel stationary units, and alcohol was experimented with as a potential new fuel, not only in Coventry, but at Cannstatt as well, the two company's keeping each other in touch with developments.

Trouble from within

In the meantime, trouble in the Boardroom was breaking up the Company from within. There were widely separated views on policy amongst the Directors, and for some time, the Company hovered on the brink of insolvency as a consequence. The root of the problem was the lack of a carefully selected Managing Director (or other chief executive), responsible to the Board for the Company's welfare. The Directors themselves tried to carry this out by devoting a couple of hours a day within their busy schedules, but with other interests at stake, they could not give their undivided attention to the matter in hand.

The differences of opinion among the Directors became acute - the Board was a house divided against itself, and an individual at war with another thinks more of his own interest than what is best for his charge. This unhealthy atmosphere culminated with a meeting of the Board on the 21st of July 1898. The sultry mood of the Boardroom finally exploded when the anti-Sturmey section (in this case, Mace and Holt) proposed that an individual by the name of Richard Bannister should be appointed as the new Chairman. Bannister, on his own admission did not know an exhaust valve from a carburettor, neither did he possess any outstanding qualities to justify his occupying such a high office at this critical time - in actual fact, he was not even a shareholder. All this caused Sturmey to see red. He lodged *"a strong and emphatic protest"* and when, after all he could do and say, Bannister was elected a Director (a position he held for less than three weeks incidentally, despite later attempts to get back onto the Board), Sturmey left the meeting with the words that *"he intended to act as he saw fit."* He did so and on the very day of the meeting, addressed a letter regarding the election of Bannister and the conduct of the anti-Sturmey party to each and every shareholder. As soon as the Board learnt of this, the feelings of the anti-

Sturmey faction ran higher than ever, and an Extraordinary General Meeting was called to consider the possible appointment of a committee of investigation in relation to Sturmey's circulars.

After J.H. Mace, as Acting Chairman, had stated the reason for calling the meeting, one particular shareholder asked why a committee of investigation was necessary at all in view of the optimistic tone at the last AGM. A lengthy discussion followed, and eventually three gentlemen - Avery, Rawlingson and Leake, all shareholders, were nominated to serve as a committee to investigate management errors. The members wasted no time in getting to work, employing professional assistance to help out. Day after day they laboured, either at Coventry or at the London showrooms. Then, on the 10th of December, they issued a long and private report to the shareholders which constituted a serious indictment of the Directors' policy.

The Reports findings

The principal conclusions outlined in the report showed that to elect Bannister as Chairman was out of the question due to his lack of experience and qualifications. However, the report went further than this, much further in fact, stating that the whole Board was incompetent - a statement which caused the deepest indignation in the ranks of both the Sturmey and anti-Sturmey parties. Another charge was that the original Directors had been nominated by the British Motor Syndicate Limited which meant that, as this concern were licensers of the Daimler patents, the interests of the Daimler Motor Company were not independently represented when the terms for the licence were made. *"We are of the opinion"* the report went on, *"that the sum of forty-thousand pounds was excessive."*

The Committee also considered that the original prospectus was misleading, in that there was no mention of the fact that the promoters retained the right to nominate two directors on the Board of the Daimler Motor Company. These were Bradshaw and Lawson, although neither their names nor the clause appeared on the circular. In language that left no room for misinterpretation, it was stated that no member of the Board had had the technical experience which was so essential for the well-being of the business, adding that *"The Board appears to have divided itself into two parties, and the differences of opinion have been frequent and acute. The result has been that the Board has not exercised that control over the expenditure of the Company's funds which the shareholders have a right to expect; important matters have been constantly deferred, and the business of the Company has not been conducted in the efficient and energetic manner which is so indispensable for the success of any commercial enterprise."*

Afterwards, the profit and loss and balance sheets that had been presented at the first AGM were criticised. What had been shown as a very satisfactory and substantial balance, should in fact have showed a rather large deficit after such purely unexpected windfalls which were in no way connected with the normal trading of the business (such as the sale of the building to the Great Horseless Carriage Company), had been taken into account.Further complaints were that the number of the Directors was excessive, so too were their fees, and the granting of advantageous terms for the purchase and maintenance of directors' vehicles was quite improper.

Summing up, it was noted that the Company was only able to continue trading due to the timely financial assistance of the Hon. Evelyn Ellis but, as that sum had already been absorbed, a very large amount of additional capital would be necessary if the business was to be continued. Despite these many problems, the Committee concluded their report by emphasising their belief in the future of the new industry, and that an efficient management team with the command of a fair working capital would make the Company a successful undertaking. They strongly recommended that a paid Managing Director with a substantial interest in the Company should be appointed, and that the Board should consist of not more than five and not less than three Directors, who should be chosen and paid at a rate fixed by the shareholders.

This merciless document had the effect of uniting the Board at least for a short time (they even agreed to engage Sidney Straker to help deal with the report), but only for the hopeless purpose of finding a suitable reply to the Committee's findings.

Their reply came on the third day of 1899 - they certainly set up a tolerably strong defence on some points, but were painfully weak on others. Four out of the seven Directors were retiring anyway, and the Board had no objection to the shareholders appointing a Managing Director, but they deplored the very suggestion that the Company should be wound up, even for reconstruction. In their view, it would be an awful waste of money; the debenture holders interests would inevitably arise, and it was hoped that at the forthcoming AGM, when all the Company's matters would be discussed, and a vote of confidence in the

Board would be passed by the shareholders.

Keen regret was expressed at the way in which the Committee of Investigation had issued their report. Not only had the value of Daimler shares dropped quite substantially, but it had also succeeded in lowering the prestige of the Company name. Gottlieb Daimler and J.A. Bradshaw had both resigned their seats on the Board, and the Hon. Evelyn Ellis and William Wright retired by rotation - neither offered themselves for re-election.

If the Company was wound up (and the Committee seemed to be doing all it could to achieve this), Henry Sturmey would have seen his £4,000 of Daimler shares disappear before his eyes, in view of Ellis's debentures. The other Directors too, were in a similar boat, but no personal consideration would move the Committee tochange their stance from making a new and fresh start. Through their solicitors, the Committee had expressed their wish to take control of the business.

A stormy AGM

At the AGM, held on the 7th of January 1899, it was stated that sales had increased, resulting in a healthy profit of some £9,381 for the year, and that Daimler had won the £100 first prize at the Royal Agricultural Society's motor trials in the light car class. The Admiralty had approached the Company with regard to launch engines, and even foreign firms such as Charron were buying components from them. However, this could do absolutely nothing to stop tempers running high. It was a stormy meeting right from the start, which was not helped by J.H. Mace's bitter comments on the Committee's report - this angry scene continued for no less than three and a half hours. It was suggested that another committee should be appointed to investigate thoroughly the financial situation of the Company as it stood then, and that it be asked to submit a definite proposal for putting the business back on a sound footing within a month.

Eventually, votes were taken, resulting in a victory for the remaining Directors with a majority of over 2,000 votes. Sir Edward Jenkinson and Edward Bayley were elected to fill the vacancies on the Board, with the former being appointed as Chairman. So ended one of the first storms that the Daimler Motor Company had to weather in its passage to calmer waters; that Daimler ever survived into the next century at all is one of the many wonders of the British motor industry.

Below: *An 1898 Daimler.*

CHAPTER TWO

A FRESH START

The events covered in the last chapter, in particular the retirement of several of the original Company Directors, were followed by even more sweeping changes in staff. A.H.D. Altree resigned, followed a short time later by Ernest Instone, although the latter would always be linked to the Daimler marque thanks to his formation of Stratstones in London. A drastic reshuffle of the clerical and official staff was carried out in London resulting in almost all of them being dismissed, and saving the Company over £500 a year. George Foster Pedley, who had previously been with the London Van & Wagon Company, was given the post of London Manager, whilst Percy Richardson joined the staff as Daimler's representative after leaving the Great Horseless Carriage Company. Van Toll left during the summer of 1899 to establish an Agency in London, and Sidney Straker became the Company's Consulting Engineer, retained at £350 a year.

These many changes were in fulfilment of promises given by Bayley and Jenkinson to carry out a thorough and exhaustive investigation into the Company's affairs. Gradually, by revising book keeping methods, closing the works on a Monday, and cutting back on unnecessary staff at Coventry (nearly two thirds of the production staff at the Daimler Works were dismissed), things began to look up. Profits were far better, and a new-found energy was shown by everyone involved.

Facing page: *The first Royal car, delivered in 1900. Seen here, now fully restored, in the grounds of Sandringham House.*

A7

Above: *The Critchley Light Car.*

During the last few months, Sturmey's position on the Board had become a rather embarrassing one. He had ceased to take any active part in the Company's business and so, during the May of 1899, he offered his resignation. Addressed to the shareholders of the Daimler Motor Company, he wrote the following:

Dear Sir or Madam,

With regard to the above Company, I regret to have to inform you that a line of policy has been adopted which speedily reduced our affairs to a chaotic condition, and, although some slight efforts have since been made to amend matters by a reversal of policy in a few particulars, it is plain my experience can be of no further service to the Company, as I object to being a "director" of a business I have practically no voice in the direction of, or, as a director, to be responsible for the results of a policy diametrically opposed to my views, I place my resignation in your hands.

Yours Faithfully,

Henry Sturmey.

Doubtless this was a great relief to the other Directors of the Company, as Sturmey had become little more than a thorn in their side by this time, even taking up posts with potentially rival firms. He had also used the pages of *Autocar* magazine to voice opinions that really should have remained private - it was a sad and bitter end that could easily have been avoided. Sturmey went on to be fired by Iliffe too, when eventually the famous publisher decided that Sturmey had used his position as Editor wrongly. It should be remembered, however, that Henry Sturmey did do a lot for the cycling and motoring cause; he died in 1930 at the age of 72.

The Critchley Light Car

While all this was happening, important events were taking place at the works. For some time, around fifty twin cylinder 3hp (4 bhp) water cooled engines had been in storage. They had been sent over from Cannstatt, but were no longer suitable for any standard vehicles likely to be produced in Coventry; Critchley suggested that he should be allowed to design a light car to make use of them.

After a certain amount of hesitation due to the vehicles quite unorthodox layout, the Board eventually gave the go-ahead, and the "Critchley" light car was born. Whatever shortcomings this car possessed, it should be remembered that its sole purpose was to use up surplus engines and, had that necessity not arisen, it is unlikely that it would have seen the light of day.

Simplicity of construction was the main feature; the twin cylinder engine was set across the frame so that the flywheel rotated in the direction of travel, thus allowing it to be used as a pulley for the high speed belt. There were no jockey pulleys - a belt guide served to run one or the other belts from a loose pulley to a fixed one. On a countershaft were corresponding pulleys, and in the centre of this shaft was a small hide pinion which meshed with a large bronze crown-wheel to the differential in the usual fashion.

The first few models lacked any rear springing, although this situation was altered in the later examples. Looking at the photograph of the Critchley, the curious position of the steering wheel will be observed, and the lever to the left of the driver which operated the gearchange by moving each belt as required. Only one brake was provided, a foot brake which came into contact with the wide rim of the high-speed pulley on the countershaft. In July, it was announced that the first twenty had already been sold, and contemporary reports were generally flattering.

The RAC's first Motor Show

It was during the summer of 1899 that the recently formed Royal Automobile Club organised its first show. It took place at the Old Deer Park in Richmond, and involved a week of exhibitions and demonstrations. Beginning on the 17th June, it was preceded by the country's first ever hill-climbing event at Petersham Hill. Running from the picturesque village of Petersham, competitors had to negotiate a 1:9 gradient before arriving at the finish signified by the Richmond Park gates.

There was also a long distance trial starting from the Red Lion Hotel in Southall, and covering around fifty miles before finishing in the same place.

Of course, 1 in 9 gradients and "long distances" of fifty miles do nothing to worry the driver of today, but it should be borne in mind that in 1899, these were quite testing objectives.

The Daimler entry was a fairly substantial one, with two Critchley machines, a Siamese Phaeton, and a Wagonette. In addition, there was a twenty mile trial on the following day, in which Critchley himself drove a 12hp Daimler parcels van (at that time, the largest van of its kind to have been constructed in the UK, and one of three built for the Post Office). The judges awarded Gold Medals to the Critchley light car, the Siamese Phaeton, and to the Rougemont Wagonette - an excellent result for the Company. Afterwards, the RAC released the following statement:

"The Daimler Company exhibited a large number of Daimler vehicles with various forms of bodies. Among these were the Siamese Phaeton, the Rougemont Wagonette and the "Critchley" light Daimler, all of which were entered for and went through the hill-climbing and the fifty miles efficiency tests. They also exhibited a Mail Van which was entered for and passed through the heavy vehicle trials.

"All these vehicles were fitted with standard forms of Daimler motor, and two of them had new forms of gear in place of the standard Daimler pattern, as in the Rougemont Wagonette and the Mail Van.

"The new feature of the Siamese Phaeton was the change speed gear, the details of which are described in the

Below: *Three of these four cylinder, 12hp Daimler vans were supplied to the post office in mid-1899. At first, cooling presented a few problems, but this was later remedied by fitting the radiator on the roof.*

Above: *King Edward VII (then Prince of Wales), with Lord Montagu of Beaulieu in August 1899. The 12hp Daimler is a very famous machine. It was the first British built car ever to race on the Continent.*

Automotor for June. This gear will have undoubted advantages over the gear of the older form, the teeth of the wheels always remaining engaged, the one or the other speed being obtained by clutching one of the four which are normally loose on the shaft. The wear which usually takes place in the gear teeth is thus transferred to much less costly and easily renewable parts, and greater smoothness of running is secured. Minor improvements were also shown in the lubricating arrangements and in the steering gear. In general arrangement these cars were the same as those which proved satisfactory in the club tours and on occasions when under the observation of the judges."

A large number of distinguished visitors attended the events. HRH Prince Edward of Saxe-Weimer declared the show open, and demonstrations began immediately after lunch. The actual exhibition of vehicles was held in the various buildings scattered throughout the grounds, whilst the lawns were used for demonstration runs.

Incidentally, a four cylinder 12hp Daimler that the Hon. John Scott-Montagu purchased in 1899, has been rightfully preserved for future generations in the National Motor Museum. Not only was it the first petrol engined machine to enter the Houses of Parliament, but also took part in the Paris-Ostend road race held in the September of that year. Although not highly placed (it took third in the Touring Class), it became the first ever British built motor car to be raced by an Englishman on the Continent, and is perhaps best known for the photograph of Beaulieu and the Prince of Wales, taken in August 1899.

In November 1899, the old Crystal Palace played host to the National Cycle Show. The National Show and the Stanley Show (held in Islington) were still the two main exhibitions for manufacturers to display their new models, and it was the National Show that held the advantage of the Crystal Palace grounds.

On this occasion, the Daimler Company showed various vehicles, including the popular 6hp models with numerous body styles. It was recognised that radiators, or condensers as they were often known in the early days, were becoming a necessary item, and the Daimler range now included this feature as a standard fitting, Cannstatt having led the way. It was also apparent that pneumatic tyres were superseding solid rubber ones; pressure fed engines were taking over from those with gravity fed lubrication, and tiller steering was gradually being replaced by the steering wheel on most of the cars - it first being discussed in the final months of 1898.

Daimler's Achievements

The Company took every opportunity to publicly prove their products, and it is to their credit that they achieved such a great deal of success prior to the turn of the century. The following is a list of awards gathered by Daimler between 1898 and 1900.

Above: *A new style of Daimler Phaeton, introduced during the November of 1898. The car is in fact much higher than it looks in this picture.*

Gold Medals:
Liverpool Cycle & Motor Show, 1898.
For design, construction and efficiency of motor carriages.
Liverpool Cycle & Motor Show, 1899.
For design, construction and efficiency of motor carriages.
Awarded by the Automobile Club of Great Britain & Ireland, June 1899.
For design, construction and efficiency of the Daimler cars.
Awarded by the Motor Car Journal.
For efficiency tests of a Daimler Victoria, which went through the efficiency runs of seventy miles for three consecutive days without a single stoppage carrying a full compliment of passengers.
Awarded by the Motor Car Journal.
In connection with an efficiency test held during July 1899 in respect of a Daimler parcels van. This van went through the efficiency runs of fifty miles per day for three consecutive days carrying a load of 15cwt without a single stoppage.

Silver Medals:
Awarded by the Automobile Club of Great Britain & Ireland, June 1899.
For design, construction and efficiency of a 12hp Goods Van.
Highest award by the Highland & Agricultural Society.
For the best motor vehicle exhibited at the Perth Show, 1898.
Highest award by the Royal Lancashire Agricultural Society.
For a motor vehicle exhibited at the Blackburn Show, 1898.
Awarded by the Ormskirk & Southport Agricultural Society.
For the best motor car exhibited at their show, 1898.

Bronze Medal:
Awarded by the Automobile Club of Great Britain & Ireland.
For improved braking for motor vehicles, June 1899.
In addition to these, there was also a £100 first prize for a Daimler van which won the Royal Agricultural Society of England's "Trial for Self Moving Vehicles", held in Birmingham during the June of 1898. More success came in the following autumn, when a 6hp Daimler Universal Dog Cart completed a 100 mile journey along the Oxford Road without a stop.

The Third AGM
The third AGM of the Company was held on the 18th December 1899 and, in view of all the changes on the Board and staff that had taken place, the effects on the fortunes of the Company were keenly awaited. Unfortunately, Daimler had been made responsible

Above: *Daimler No.13 on the famous 1000 Miles Trial of 1900.*

for settling a number of claims involving accidents and breakdowns, including Britain's first ever fatal motoring accident. Car insurance was practically unknown in those days, so the substantial damages involved had to be paid out of Company funds. Furthermore, the old Board had made a bad decision regarding the purchase of a launch named "Daimler" which had been moored on the Clyde. It had cost over £1,000 and was later destroyed by fire; although this was recovered through insurance, the wages of the men operating the launch amounted to something like £680. All of these matters were explained in detail to the many shareholders present.

It was also noted, however, that orders were still flooding in, and altogether the outlook was much brighter than before. The Chairman had to report the resignation of Sir Edward Jenkinson for reasons of ill health, and the promotion of J.S. Critchley to General Manager. At least the third AGM had passed in a rather more friendly atmosphere than the last one.

Looking at the British motor industry with hindsight, it is possible to see it as a series of clearly defined areas, each one divided from the last by virtue of some influence, which at the time may not have seemed in the slightest bit significant. The gradual change in public opinion from ridicule to dislike, from dislike to acceptance and then to enthusiasm, were in themselves milestones along the path of progress. At the end of the century, the industry was highly unstable - few people outside the big city's had seen a motor car, and engineers were deeply divided in their views. Indeed, one of the very points on which almost total agreement existed was in the number of wheels, but even then there was their size to consider and whether or not they should all be of the same diameter!

This was partly the reason behind F.R. Simms' proposal for a Daimler Union. *"Mr Simms' letter of July 7th 1899 asking for an interview with the Directors of this Company to discuss a scheme for consolidating the interests of all the Company's interested in the Daimler Motor was read, and the Secretary was instructed to invite Mr Simms to attend the next Board Meeting at 2.30p.m. on the 27th inst. at 219 Shaftesbury Avenue, London, WC."* It was a good idea, but nothing came of it despite several meetings on the subject.

MMC were still buying engines (6hp units at £80, and 12hp at £190), and Count Zborowski ordered a racing car for £1,000. It would be 18 months before he saw it, but the Hon. John Scott-Montagu and Alfred Harmsworth were in exactly the same position.

In January 1900, the prices were raised on the 6hp cars. The *"ordinary standard 6hp motor and frame, with solid rubber tyres, radiator, plated fittings, accelerator, and aluminium gearcase"* would be £350, while the *"1900 Type 6hp motor and frame, with wheel steering, geared water pump, aluminium base to engine, and pneumatic tyres"* was £425 list.

The Thousand Miles Trial

At the turn of the century, public ridicule had changed to hostility, and not altogether without reason. The surface of the highways was then completely unsuited to motor vehicles, and the reputation for reliability in the earliest vehicles was not exactly a good one. What was desperately needed was some sort of demonstration which could extend through the length and breadth of the country, so that people could see for themselves that motor cars were indeed a practical form of transport.

The Automobile Club had been doing some useful propaganda work in the form of club runs and demonstrations, but their field of activity was preaching to the converted and far too restricted to reach the masses. Even at this stage, there were hundreds of miles of road over which no motor vehicle had ever passed. No-one realised this more than the late Lord Northcliffe, then Alfred Harmsworth. It was Harmsworth who first suggested the Thousand Miles Trial, and promised substantial financial assistance to get the idea off the ground.

It was eventually decided to hold this mammoth event during the spring of 1900. To say that this gave the industry the necessary boost it needed is something of an understatement, for the Trial was one of the most important motoring promotions ever held. Not only did it raise public awareness as to the automobiles potential, but it also had the effect of settling many points of design through competition. The importance of brake design and the desperate need for road improvements were all problems highlighted by the Thousand Miles Trial, prompting their rapid development.

The start took place on the 23rd April at Hyde Park Corner, with the first day taking in Maidenhead, Newbury, and Marlborough before finishing in Bristol. The route included Birmingham, Manchester, Kendal, Carlisle, Edinburgh, Newcastle, Leeds, Sheffield and Nottingham, and at most of these places, a small exhibition was held. There were also a number of hill-climbing competitions en route at Taddington Hill, Shap Fell, Dunmail Rise and Birkhill before the drivers returned back to London.

The Trial was a major triumph for Daimler. Thirteen Daimler cars started and all of them finished, with no other make of car putting up anything like the same performance. Of these, three were entered by the Company, with the others coming from private owners. All in all, the Daimlers came away with three First Prizes, five Silver Medals, three Bronze Medals, three prizes of £10 each presented by the *Daily Mail*, and two second places in Classes C and D respectively.

For all but two of the cars to receive awards of the highest calibre, is surely an achievement ranking with anything in the long history of the Daimler car, and brought its just rewards.

A Royal Car

During the early part of 1900, the Company received

Below: *An MMC on the 1000 Miles Trial.*

the honour of an order for a twin cylinder 6hp Mail Phaeton from HRH the Prince of Wales (see first picture of chapter two) who shortly after was crowned King Edward VII. His earlier brief trials had no doubt appealed to his sporting nature, and obviously, the great success of the Thousand Miles Trial had been brought to his attention. This Daimler was the first ever car to be supplied to a member of the Royal family, and what follows is part of the *Motor Car Journal* article that appeared in June 1900:

"There have been many reports current during the past few months as to the car which has been ordered, through Mr Stanton by the Prince of Wales, but none of them have been of an authoritative nature. We ourselves have naturally been cognisant of these rumours but decided to hold over reference to the Royal Car until it was a fait accompli, and until such times as we could give definite details of the actual vehicle.

"This we are able to do this week, and by the kindness of Mr G. Foster Pedley of the Daimler Motor Co. Ltd, we are in a position to publish a couple of illustrations of the car that has just been delivered to His Royal Highness. This vehicle itself, on which we have had the pleasure of a short trip, is one of the Daimler Company's standard type, the frame being that known as Model A. The motor has the usual two cylinders and is of 6hp, an innovation being the fitting of double ignition, both electrical and by means of an incandescent tube. The electric ignition is normally relied upon, the burners being kept in reserve. The adoption of double ignition necessitates but a small alteration to the motor; the sparking plugs are located in the top of the cylinder head; the induction coil adopted is one of Blake's and so far the arrangement has given complete satisfaction. A new departure in connection with the motor is that the 'accelerator' is controlled by a foot pedal in place of the hand lever hitherto employed. In the construction of the motor, aluminium has been used as far as possible, the crankcase, etc., being made of that metal. The transmission gear is of the usual Daimler pattern, giving four speeds ahead and reverse motion, and the change of speed is controlled by a hand lever at the side. The car is capable up to a maximum of about 24mph.

"A few details may now be given as regards the Body. The coachbuilding part of the work was entrusted to Messrs Hooper & Co. Ltd of St James's Street, and it is needless to add that they have carried out the same in an eminently satisfactory manner. The body takes the form of a mail phaeton, with seating accommodation for four persons. The woodwork is painted in the Royal colours, chocolate and black, picked out in red. The upholstery has been finished in colours to match and altogether the car has a quiet and stylish appearance. Beyond the painting, the only indication that the car belongs to Royalty is His Royal Highness's crest affixed to one of the panels behind the lamp on the left side. The front seat is provided with a collapsible hood; this is a new type as far as English motor cars are concerned, it being constructed on the lines of the American buggy hoods, the back being made to roll up, thus enabling the hood to be used as a canopy in bright sunny weather such has been experienced during the past week. The road wheels are of wood shod with Clipper pneumatic tyres; the mud guards or splashboards are of leather and of a much greater width than usual. Steering is controlled by an inclined hand wheel, the standard of which is pivoted in a vertical position to facilitate mounting and dismounting.

"Having briefly dealt with the leading features of the Royal car, which does credit to all concerned in its production, we may devote a little space as to its career during the past few days. His Royal Highness spent a good deal of time at Sandringham in learning the mysteries of the motor car under the guidance of Mr Stanton, on the latter's new Daimler mail phaeton. The Prince quickly learned how to manage the reins, or rather the steering wheel, and it was not long before he was driving the car on second speed in the grounds of his Norfolk residence. So pleased was the Royal chauffeur with his experience that he was eager to see his own car, which made its debut at Ascot on Monday last. The question of a reliable mechanician was a matter not settled in a day, and after much deliberation, a Mr Sidney Letzer was selected for the responsible position. Mr Letzer has been with the Daimler Motor Co. Ltd for about two-and-a-half years, and has a thorough acquaintance of the vehicle of which he will now be in charge." The article went on to outline the car's duties in its first week, but later said; *"It will gratify those who have watched with interest the progressive development of the British automobile industry to learn that the Prince is well satisfied with his car, which is British-built throughout. Seeing that such excellent and reliable vehicles can now be constructed in this country, he cannot understand why so many automobilists should show a preference for those of continental manufacture. Certainly, in this he is maintaining the traditions of the Royal family, who since the example set by the late Prince Consort have done so much to set the fashion in promoting the welfare of English trade and industry.*

"The position which the Prince occupies at the head of British society - and the esteem in which he is held on the Continent, makes his adhesion to the ranks of automobilists an event of the greatest importance, and should give an impetus not only to the industry, but also to the popularity of the motor car as a pastime and sport." Indeed, the Prince of Wales' order for a motor car gave the industry a whole new respectability. Where the Prince led, others

Above: *An interesting picture of the 'assembly hall' at the Cannstatt Daimler works in 1900.*

Below: *The Motor Mills, pictured in 1900 with a group of Daimler owners.*

Above: *More MMC machines, this time on thier way to a show in Birmingham.*

Below: *Mr Letzer, the first Royal chauffeur, with the second Daimler delivered to HRH the Prince of Wales in 1900. This model being a 12hp wagonette.*

followed, and no sooner had he bought and used his car than society accepted the new idea. Also, from that moment on, the Daimler became the Royal Car.

Sad News from Germany

In the meantime, other events which were to influence the destiny of the Company had occurred. On the 6th of March 1900, Gottlieb Daimler died as the result of heart disease. He had been a Director of the business in Coventry from when it was first formed, and although he had played no part whatsoever in its administration (a Board Minute dated 13th May 1897 excused Daimler from attending Board Meetings for six months, and two similar statements were issued before his resignation in July 1898), he was well known to most of the existing Board, not least by those who had gone to visit his works in Cannstatt.

As his invention formed the basis of the Daimler Motor Company's activities, naturally his death was keenly felt in England. The great man was buried in the Uff cemetery at Stuttgart, and his funeral was attended by a vast gathering of well known motoring people, including René Panhard and Mme. Levassor.

Then, shortly after the Thousand Miles Trial, J.S. Critchley announced his resignation to enable him to take up an appointment with the Brush Electric Traction Company. He had been with the Company since its formation, and his name will always be associated with the Daimler marque. He retained an interest in the Company until 1902, the year in which Brush Engineering and the British Electric Traction Co. began building cars. Otto Mayer, who had been with the Company from its birth, having come over from Cannstatt to help the fledgling concern, also left at the end of 1900 to further his career in the industry.

The Appointment of Percy Martin

Critchley's resignation led to several officials being given new positions, but by far the most important appointment made during this period was that of Percy Martin, who replaced Critchley. Martin was born in Colombus, Ohio, in 1871, and was one of the very few American's to have any bearing on the British motor industry during those early days. After acquiring his engineering experience in the United States, Martin left America in 1894 to join the Union Elektricitaets Gesellschaft in Berlin, remaining there until he left for Milan to open a new factory for the same concern in Italy. A financial crisis forced the company to close down their Milan branch, whereupon Martin decided to return to the States. However, just by chance, he broke his journey in London, and it was here that H.F. Orcutt (who had been with the Company for some time) told him of the vacancy at the Daimler works in Coventry. After starting as Works Manager on 23rd October 1900, he continued with the Company for 35 years - long enough to see the business develop into one of the country's finest marques.

The 1900 Motor Show

At the 1900 National Show, the Daimler Motor Company exhibited three 6hp cars, two of which were fitted with both electric and tube ignition. A 16hp car built to the order of J.D. Siddeley (of Armstrong-Siddeley fame and later Lord Kenilworth) was also shown, as was a "Critchley", although this particular machine was one of the slightly redesigned versions which had been renamed the "Kimberley".

The Kimberley was much the same as its predecessor, but the belt-drive was modified by Straker so that three gears and just one belt was used, as this had been the cause of embarrassment for several owners. Claude Johnson had used a Critchley to recce the Thousand Miles Trial, but found that belt slip was a major problem on steep hills. Indeed, the only way to get the belts tight enough to transmit drive in these extreme circumstances induced the engine bearings to overload and fail! It is interesting to note that the Company had thought about a 2hp car as well (also belt driven) but this was not carried through. The engines - probably from Bollée of Le Mans - were later sold to a Mr L. Epstein.

In addition to the Kimberley, there was what was rather ambitiously termed a 20hp racing car, shown as a chassis. Whether or not this was the one being built for Count Zborowski is not known, however the dates do correspond. This vehicle had two water pumps driven from the flywheel, and a radiator immediately in front of the engine crankcase. An interesting feature of the engine was the facility to release the compression on three of the four cylinders to aid starting.

A new charabanc, designed by Sidney Straker, also had a number of novel features. Instead of using chains and sprockets to transmit the final drive, he opted to use a method favoured on the Cannstatt Daimlers, namely having pinions mounted on the half shafts that meshed directly with internally toothed wheels attached to the road wheels. In this, the second Daimler bus (Critchley had demonstrated a 16 seater omnibus as early as October 1898, and sold it for £750 to the London Steam Omnibus Co.) the standard 12hp engine was retained, and a 26 seater double-decker followed soon after, resulting in the first double-decker bus service in the City of London.

A new four cylinder 12hp car was also introduceed, which represented a distinct step forward in design. Its comparatively modern lines, drawn up by Percy Martin, and its well-raked steering column showed a definite break from previous practice.

More Royal Cars

Two more Daimlers were ordered by HRH the Prince of Wales, taking his total to three during that one year. His original car was returned to the factory, but the new cars were a 12hp four cylinder wagonette for his own personal use (£870), and what was termed as a Beater's Car, designed to carry as many as 14 people during the shooting season. This car, which cost £975, was used extensively by the Royal household, and by guests attending the various shoots held at Sandringham, Windsor and elsewhere.

1901 Types

It was noted that the cars for 1901 should have the following features:

Engine bonnet all aluminium if possible.
Radiator in front of engine bonnet.
Pressure lubricator, wherever practicable.
Ignition, both electric and tube.
Indicator gauge to water circulation.
Cushioned steering gear with new worm.
Gravity pattern carburettor below burners.
Sprag, to be tube if practicable.
Sprag to be fitted to front axle, if found successful.
Motor to be two-cylinder, developing not less than 7.5hp.
or four-cylinder developing not less than 16 hp.
High compression to be used, if found satisfactory.
Weight of two-cylinder frame not to exceed 17.5 cwt.
Weight of four-cylinder frame not to exceed 20 cwt.
Length behind dashboard:-
Two cylinder car 6' 8".
Four cylinder car 6' 4" and 7' 4".
Gearcase; two cylinder to have present 12hp gearcase.
Four cylinder to have new twin gearcase - both to be adapted for roller bearings.
Axles to be of Lemoine type.
Brake pulley to be water cooled.
Band brake to act both forwards and backwards.
Gear wheels, width to be increased.
A throttling arrangement to be used if found satisfactory.

In the end, the 1901 range was as follows:

6 hp	*A Frame*	*£390*
6 hp	*B Frame*	*£320*
8 hp	*Frame (roller bearings)*	*£500*
8 hp	*Frame (plain bearings)*	*£445*
12 hp	*Frame*	*£735*
16 hp	*Frame (roller bearings)*	*£1,000*
24 hp	*Frame (roller bearings)*	*£1,350*
Kimberley	*Complete*	*£225*
8 hp Van	*Frame*	*£415*

An unexpected offer

During 1900, the fortunes of the Company had began to show signs of improvement. At the fourth AGM, the Chairman reported that the position of the Company had steadily improved despite the fact that already many other motor manufacturers had fallen by the wayside over the past year. However, during February 1901, both E.H. Bayley and Sherwin Holt resigned from the Board. The Chairman announced that he too wished to resign, but was prepared to continue for a time in order to avoid the absurd situation of the Company having just J.H. Mace on the Board. In fact, Mace had previously stated that if Bayley and Holt resigned, he would follow them, but at the last minute he changed his mind, presumably out of consideration for the welfare of the Company. For the leading car producer in England suddenly to lose its entire Board of Directors would inevitably have an adverse effect on the value of its shares. It was while matters were in this uncertain state that Harry Lawson saw his chance.

It will be remembered that it was Lawson who had originally promoted the Company, and for a short time, was the Daimler Chairman, before retiring in late-1897. Although he had remained a shareholder, he had taken no active part in the administration of the Company and indeed, little was heard of him until he made a remarkable offer to purchase the business.

Briefly, Lawson's offer was that £60,000 should be paid in cash to the shareholders, and preference shares issued to the value of £50,000, giving them 11 shares for every 10 they owned. It was a good offer, but naturally, the remaining Directors could neither accept or refuse without first consulting the shareholders. A small committee was established to look into the deal, and approval was given, but for the minor point of Lawson's guarantee; he had already raised his original offer to £5,000, but when asked to double it, he immediately broke off negotiations.

Above: *A private omnibus built on the Daimler 12hp four cylinder chassis, dating from around 1901.*

Below: *A 1901 16hp Daimler being driven by Countess De Carrie, Prima Donna at La Scala, Milan. In the passenger seat is the famous automobilist, Montague Grahame-White.*

Above: *Oliver Stanton at the wheel of "Le Chat Noir", a 1902 22hp four cylinder Daimler.*

There was disappointment in some quarters, and an EGM was called to openly discuss the Lawson matter and elect a new Director. On the subject of Lawson's offer, Sherwin Holt in particular criticised the Board for holding out for a larger guarantee. In view of the value of Daimler shares at the time, he felt £10,000 was totally unreasonable, and didn't blame Lawson at all for pulling out. In the author's opinion, this was perhaps a lost opportunity for shareholders to make some fast money, but long term, it may have been a blessing in disguise. Although one has to admire Lawson for his vision, he was perhaps not the best man to head the business; his reputation alone would have been enough to wipe vast sums off the shares of an already ailing concern!

Little was heard of Lawson after this. His plan for owning a *"master patent"* had fallen through, and a number of claims he made regarding patent infringements failed. After serving twelve months hard labour for making false statements, he was a changed man. He died in 1925, leaving less than £100 from what had been a massive fortune.

Meanwhile, Sir Edward Jenkinson's re-election was proposed, but he stated that he would only accept on the condition that he, and the other members of the Board, had the right to appoint Directors if it was in the interest of the Company. He was granted his wish, and by April 1901 he had become Daimler's Chairman again. It became apparent to him that the whole policy followed beforehand was in need of drastic changes if the Company was to survive. The most pressing reforms were the need for a new Board, a Managing Director, and some fresh working capital.

J.D. Siddeley approached the Company to, see if they would support his entry on the Paris - Berlin Race, and indeed, Daimler gave him a riding mechanic to look after his 16hp car. However, shortly after his request, the Board asked Siddeley if he would consider becoming the Managing Director of the Company. He declined the offer on the grounds that he was just establishing his own business, but the link between John Davenport Siddeley and Daimler remained.

Daimler Motor Manufacturing Co.

With all this uncertainty, the thought had crossed the Board's mind of a possible merger with MMC. In June 1901, *"The financial position and future prospects of the Company were explained by the Chairman. The Chairman was authorised to go on with the question of amalgamation with the Motor Manufacturing Co., and report further to the Board."* A joint report was commissioned, and after some progress towards forming the "Daimler Motor Manufacturing Co." it seems as if the MMC were the ones who stopped the merger.

At this time, they had debts of £6,000, but it must be said, Daimler were hardly in a strong position

Left:
A 1903 Daimler model with "Mayfair" style covered phaeton body.

Left:
A similar chassis to that above but this time bodied as a "Duchess" phaeton.

Left:
A 1903 22hp Daimler tonneau.

Left:
This Hooper bodied 22hp Daimler was delivered to the Prince of Wales in July 1903.

themselves at this stage. Daimler made a fresh offer in January 1902, but that too was rejected. Ironically, the tables were turned after this, and soon Henry Sturmey was writing to the Daimler Board on behalf of the MMC for a possible amalgamation *"on any terms"* - nothing came of it.

A worthy leader
With Jenkinson's reforms in mind, another meeting was held to ask the shareholders if they would sanction the issue of new debenture or preference shares to raise another £35,000. Sir Edward stated that this capital was urgently needed, and would be used to pay off the existing debentures held by the Hon. Evelyn Ellis, as well as give the Company the necessary freedom to exercise a new policy of steadily building up a stock of cars rather than build to order. After some discussion, the issue was approved, and towards the end of February 1902, the public prospectus was released.

In the meantime, the Company was engaged in some rather curious litigation. The original British Motor Syndicate Ltd had been reconstructed as the British Motor Traction Company Ltd, and during the summer of 1901, this new concern wished to register a business called the Daimler Wagon Company Ltd. Because of the similarity of the names, Daimler sought an injunction to restrain them from doing so, on the grounds that the two firms would be confused. Evidence was given by a number of pioneering motorists, including J.S. Critchley, the Hon. John Scott-Montagu, S.F. Edge and several others. The outcome was a somewhat surprising one, with the Judge stating that the title had not been calculated to deceive the public, because there were already many firms using Daimler engines in their products, and others which manufactured motors of the same type. The eventual outcome was that the action was dismissed with costs.

The Glasgow Trials
It was during the September of 1901, that the 500 Miles Glasgow Motor Trials were held in conjunction with the Glasgow Motor Show. It was the only large-scale event that the Automobile Club had organised that year, with five set routes radiating each day from Glasgow, but devised so that at least two hill-climbing tests were included. The event began on the 2nd September, and entries were accepted from both manufacturers and their agents, and from private individuals - a similar arrangement to that employed in the Thousand Miles Trial of the previous year.

The Daimler entries were interesting, as the Company entered two 6hp light cars which were still of a largely untried new design. Their general layout was unusual as they were the first Daimler cars to employ a propshaft to transmit the drive. The chassis was of tubular construction, powered by a twin cylinder engine equipped with both tube and electric ignition, in front of which was placed the water tank and radiator. The drive was via a single-cone clutch to a three-speed gearbox, and then to a live axle through the aforementioned propshaft. Unfortunately, both cars proved to be unsatisfactory. On the day before the trials, a brake test led to car No. 6 stripping the teeth off its bevel pinion; it was withdrawn. Then, on the second day, the other 6 hp car stripped the teeth on one of its differential pinions whilst backing out of the station yard. Needless to say, the design was abandoned at once due to this poor showing.

Car No. 7 was also a new model and, at that time, it was one of the most powerful road cars ever turned out by a British factory. Both tube and electric ignition were again fitted, and the exhaust valves were positioned on the opposite side of the engine to the inlet, allowing the use of a shorter exhaust pipe. The general running of this car was said to be fair, and would doubtless have been better had the water circulation system not become blocked. The three private entries consisted of an early 4hp twin cylinder car which, considering its age, did exceptionally well, and two 12hp four cylinder machines. In all, compared to the brilliant performance put up by the Company's products in the Thousand Miles Trial, it was a pretty poor effort.

More staff changes

During 1901 and 1902, there were a number of important changes in the staff. In the February of 1901, Sidney Straker, who had been Daimler's Consulting Engineer, posted his resignation, and H.W. Bamber took his place. Bamber held this new position for just 14 months, as J.S. Critchley managed to persuade him to move to the Brush Company where he was working on a new car.

In September, M.H. Proctor became the Assistant Works Manager, and E.W. Lewis (a name well known in Coventry's motoring circles) briefly took the title of Chief Engineer before moving to the Deasy concern in 1906. George Foster Pedley resigned, and Ernest Instone, who had rejoined the Company, took up the newly created role of Business Manager.

Soon after the new Board had taken office, there were some striking changes made in general policy, and there was more capital at the disposal of the Company thanks to the success of the debenture issue. In addition, a well equipped London repair depot was opened at the Brownlow Mews in Guildford Street, where extensive work could be undertaken by the Coventry trained staff.

The 22hp Daimler

Much redesigning had gone on at the Coventry works, resulting in some fine new motor cars. The most memorable of these was the four cylinder 22hp model, famous for its elegant lines and smooth, refined running. Making its debut on the anniversary run from London to Southsea in November 1901, it is said to have caused a mild sensation, and was the work of Percy Martin and Edmund Lewis.

Below: *A 1905 28hp Daimler.*

The most prominent sponsor of this particular car was a certain Oliver Stanton. He had made his name when society took to cycling, and had given lessons in Battersea Park, teaching well known figures how to ride. When motor cars came into vogue, he saw the possibilities of acting as an independent liaison officer between the purchaser and the factory, and as such played a role of some importance in the first car purchased by King Edward VII. As mentioned earlier, Stanton gave the King his first driving lessons in the grounds of Sandringham, and for a while at least, he acted as a consultant whenever the King considered buying a new Daimler.

Stanton, it has to be said, was rather an odd character. He could often be seen at the early Motor Shows, immaculately dressed and strolling from one stand to the next, smoking a cigarette through a very long holder. His general appearance gave the impression that he was a very wealthy man, but he would never miss an opportunity at making money. For a while, he devoted all of his energies promoting new Daimler products (taking a 5% commission) and christened his latest black car, loaned by the factory, "Le Chat Noir".

During the autumn of 1901, King Edward VII ordered one of the 22hp cars and, after consultation with Stanton, the manufacturers adopted various suggestions made by him. The car, though not quite complete at this stage, was exhibited at the motor show held at Islington in the spring, after which, it was shipped back to Coventry for finishing. While it was at the factory, a detailed inspection of the car was made by Lord Suffield, and pictures of it appeared in the *Autocar* issues of late-June and early-July 1902.

It was at about this time that Daimler were looking at "buying in" a small car from the Continent. A Darracq was examined at the Works, and a "Chainless" motor car had been bought in the summer of 1901. The latter was a high powered Buchet-engined machine built in Paris from 1900 to 1903. The frame was to be used for a new 10hp car, but it was never built. The Lanchester ignition system was also being investigated, but it was not employed on the grounds that the Daimler engine required too many alterations to adopt it. However, *"At the same time, the Board agrees with their* [Lanchester's] *idea of standardising the different parts of the Motor Car, and will keep this matter in mind."*

A quiet AGM

For the first time since its formation, the Company was being run by a far sighted and enterprising Board, and although a trading loss of nearly £20,000 was reported, not a voice was raised against either the Directors or the manner in which the business was being conducted.

Before re-organisation could be attempted, it had been necessary to spend a large amount of money on improving and modifying the works at Coventry. The Board was also quick to realise that the vehicles they were selling were in desperate need of development to keep them in line with competition, especially from

Above: *F.R. Simms was fascinated by machines of war - this is his "Simms War Car" of 1902, which was powered by a four cylinder 16hp Cannstatt Daimler engine. At least it was better than his "War Quadricycle", which it has to be said, for a designer of his pedigree, was something of a joke.*

Above: *Two beautifully preserved machines. To the left of the picture, a 1903 Daimler and on the right, a prototype four cylinder Lanchester from the following year. Both machines survive and are kept as part of the Jaguar-Daimler Heritage Trust collection.*

the Continent, and this too had cost vast sums, but was undoubtedly an excellent investment in the future of the Daimler marque - the Company could not possibly have continued at the top of its field with the heavy, unreliable cars of the past, and a replacement for the new 6hp model was a matter of urgency.

At last, the Board under Sir Edward Jenkinson, had taken the bull by the horns and faced its problems. The success of the new 12hp and 22hp models designed by Percy Martin, combined with the strength of the new management team, did much to enhance the Company's reputation. The decision to back the Brownlow Mews Service Depot venture had paid off, and good profits were being shown in its first year. At the Motor Mills (or Daimler Works) there were orders for no less than 84 cars, the nett value of which amounted to around £62,000 - three times the amount recorded at the same time in the previous year.

Types for 1903

Much consideration had been given to the 1903 range. In mid-1902, a new 30hp car using the 22hp models frame and an improved transmission had been authorised, as had a 16hp car using the 12hp frame. However, both were cancelled in September that year. The question of a cheap 8hp Light Car was raised once again, but after an engine had been built, this too was cancelled in favour of a "10hp or 12 hp car" in the £550 price range. In the end, the existing 12hp and 22hp models were continued, along with a new 14hp car, and a 6hp light four-seater tonneau. At Crystal Palace, a TC.22 and TB.14 chassis was shown, as was a Daimler-bodied four-seater tonneau on the TC.22 chassis, a Mulliner limousine on the TB.22, a TB.22 with a "Rothschild" body, and a standard TA.12.

Chapter Three

THE DAIMLER MOTOR COMPANY (1904) LTD.

After 1903, the Company underwent a revolutionary change for the better. Now led by a far sighted Board of businessmen, the Daimler concern no longer had to rely on limited working capital and improvised methods to keep it afloat. It now occupied the premier position in the British motor industry, and the key men behind this complete metamorphosis were Sir Edward Jenkinson, Percy Martin, and Ernest Instone.

They had overcome a great many of problems; the debenture issue suggested by Sir Edward had in fact yielded very little, so working funds were still acutely limited, meaning that the Company still had to rely on customers handing over 33% of the purchase price of their car, then waiting several months before they could take delivery.

Other problems included the Motor Car Act of 1903, for although the speed limit was raised from 12mph to 20mph, anyone caught exceeding this was severely punished. Described as *"a disgrace to an engineering country"*, Police traps were very common, and many people were put off the idea of running more powerful vehicles because of the risk. Deposits on Daimlers dropped to just 10% of the previous year, while the sale of voiturettes increased steadily.

Facing page: *A new 15hp sleeve valve Daimler on test. This picture was taken from the office block in Sandy Lane, Coventry, in late 1909.*

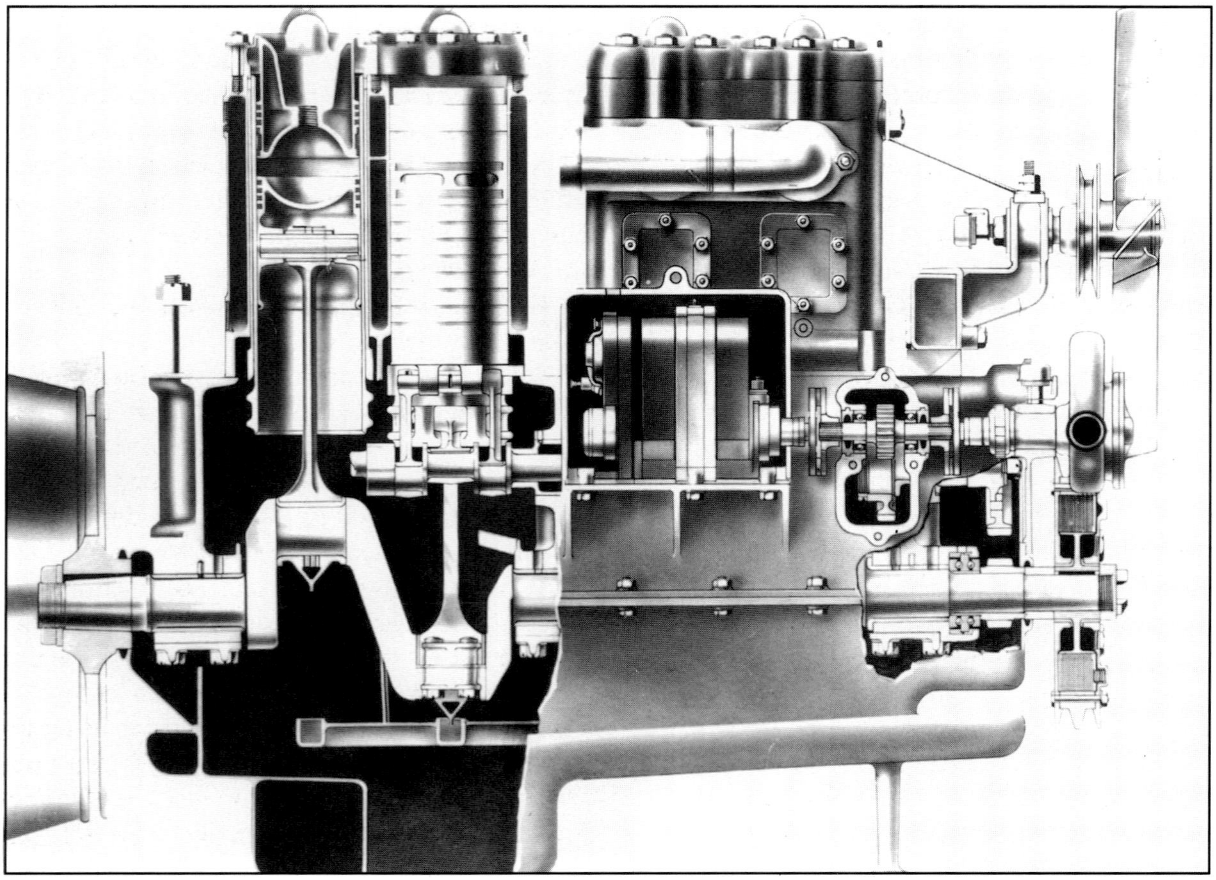

Above: *A cutaway drawing of the legendary sleeve valve engine.*

This was not an ideal situation, as the established Continental manufacturers such as Panhard & Levassor (who were producing over 1,000 cars a year during this period), Mercedes (the German Daimler) and De Dion-Bouton, could well afford to stock complete machines, thus providing their clients with vehicles straight away.

Once again the Company was prompted into considering at the possibility of "buying in" a smaller car. Société Belgica of Brussels were approached, and a contract for ten vehicles was sealed in March 1903. It is not known what types were purchased, or indeed if any of them ever made it to Coventry, but at the time, they were building two single cylinder cars, three twins, and a 20hp four cylinder machine.

There had also been a large fire at the works in the April of 1903. The Paint & Finishing Shop had been damaged, and seven cars awaiting delivery were completely destroyed; with such a limited output, this was another bad blow for the Company, and is said to have put production back for a whole year.

However, despite all of this, when the balance sheet for 1903 was disclosed, a more promising picture was presented. The nett loss from the previous year had turned into a slight profit, regardless of the costs involved in erecting a new Paint & Finishing Shop that enabled the Company to build its own bodies, and the establishment of a laboratory to check on the quality of bought in parts.

Nonetheless, there still remained the question of insufficient working capital. This lack of money was holding back all of the plans which Sir Edward knew would see the Company through harder times, and something had to be done.

Daimler in Competition

Once again, the Company had performed well in various fields of competition. The Gordon Bennett race had been held in Ireland, and this was followed by a series of short speed events in Dublin and elsewhere. Here, the Daimler's performance was better over the kilometre than any touring car, irrespective of Class. In a hillclimbing contest at Castlewellan, County Down, Daimler cars secured the first three places, and in the four mile race for the Graphic Trophy, they took three of the first four places. At

Cork, a match between a 20hp Panhard and a 22hp Daimler, resulted in a runaway victory for the Coventry car, and at Killorglin a Daimler won the Touring Class of the hillclimbing event. This was achieved without the Company themselves entering the Irish Fortnight officially, as they were *"not prepared to construct a car for this purpose."*

Public opinion may have had something to do with this, for although the Gordon Bennett was a total success, fears were running high regarding the speeds the cars were capable of. Despite being a sporting man, even King Edward VII expressed his wish for the race to be cancelled on the grounds of safety, and newspapers were highly critical of the motor car. Only after a letter from C.S. Rolls to the *Daily Mail* did things seem to calm down. In it, he pointed out that *"our hereditary instincts are shocked at seeing anything on the road faster than the horse,* [but] *cars have come to stay* [and] *will very shortly, to a very large extent, replace the horse... It is therefore no good for people to get hysterical because motor cars are driven fast."*

Meanwhile, Percy Richardson resigned from his post as London Manager, his place being taken by Undecimus Stratton, and John Gorham was given a seat on the Board after coming to the financial aid of the Company - overnight, he had become the firms largest shareholder.

Much hard work had been done during the year, and the Directors had refused to accept any fees for their services. When it was announced at the AGM, held on the 26th November 1903, that no dividend was to be paid, there was very little opposition. Sir Edward said, *"Our new types for 1904 will have embodied in them all the latest developments which tend to reliability, comfort and smooth, quiet running."*

In reality, the main item covered at this meeting was how to go about raising a substantial amount of fresh capital, and it was finally decided that a small committee should be formed to deal with this - Messrs Rawson, Avery, and Allingham were duly appointed from the ranks of the shareholders.

There were many meetings and much discussion; it had been hoped that operations could continue until the next AGM, due to be held in December 1904, but output was suffering and competition, especially from the Continent, was being increasingly felt. An EGM was called during the last week in October, and a new scheme was considered. Sir Edward Jenkinson, who presided over the meeting, pointed out that unless the Company broadened its manufacturing scope and greatly increased its

Below: *Chassis production at the Daimler works, circa 1903.*

Above: *The 1904 Bexhill Speed Trials. Mr and Mrs Stanton are in DU 47, with E.M.C. Instone in the car behind them, and Mr and Mrs Manville in DU 487.*

production with the help of more capital, the business would never make any real progress.

A new concern

Eventually, a decision was reached to form a new company to take over the existing one, with the latter then being wound up. This scheme would raise the extra cash necessary, and the proposal was carried with only four people out of all those present against the idea.

The Chairman moved that the following resolution, *"That this Company be wound up voluntarily"* was duly passed at the AGM of the Company on December 8th - the meeting was therefore brief and largely a formality, but as a result, the Daimler Motor Company (1904) Limited was formed.

While all these financial upheavals were happening, Percy Martin had been re-organising the production department of the Company. During 1904, two types of both the 28/36hp and 18/22hp models were produced, as well as the 16/20hp (the replacement for the 1903 14hp car) and a 7hp light car, which may or may not be the Belgica. The latter was a 1.8 litre twin-cylinder machine of orthodox layout, but in spite of a £300 price tag, very few were ever sold.

The largest model described above, the 28/36 hp (28 hp at 750 rpm, 36 at 1,000 rpm) had a 5.7 litre engine fitted with various ignition options, including the Eisemann high-tension magneto which was being developed at the time. The side-by-side valves were operated by a single exposed camshaft, and this model, from a design point of view, was very much the same as the smaller 18/22hp. It was very successful in competition, taking drivers like Percy Martin and Albert Farnell to very creditable results up and down the country.

The 18/22hp version had an engine capacity of just over 3.8 litres and, during 1904, one of these was ordered by King Edward VII. The limousine body for it was again designed and built by Hooper, after the King's own direction, with two revolving seats in the rear compartment for the King and a guest. The 16/20hp model replaced the successful 14 hp Daimler which had been introduced towards the latter half of 1903, and had an engine capacity of 3.3 litres.

Colin Cosier, writing in *Veteran & Vintage* in 1972, stated that he joined Daimler towards the end of 1904. *"We had a personal interest in Daimler customers of that time, since all instructions gave our customers' names and rank, and* [in this way] *we knew for whom we were working."*

The End of the MMC

No sooner had the new Company been established, than the far reaching changes in policy and production suggested by the Board were put into action. This was largely helped by the liquidation of the Motor Manufacturing Company Limited who, it will be remembered, were formed from the ashes of the Great Horseless Carriage Company.

This firm had leased their building on the Motor Mills site from Daimler, so when the latter bought back the factory, plant, and various other buildings that the MMC had erected (including a foundry in 1898), this greatly increased Daimler's production potential. There was a massive increase in floor space, and many new hands were employed, taking the total workforce at this stage to around 600 people.

By the time that the Motor Manufacturing Company sold everything to Daimler, in August 1905, it had produced a varied range of cars (usually listing three types each year), and not just Daimler-based vehicles. They had built De Dion engines under licence for use in MMC motorcycles, and indeed, by 1903 at least, the cars took on a unique identity, far removed from those built across the courtyard.

There was in fact an attempt to revive the marque in London in 1907, with a large six cylinder machine being listed. Unfortunately, only experimental prototypes were to appear from the Clapham factory, and the name was later used for a

Below: *The Daimler Works - the Motor Mills - in around 1903/04.*

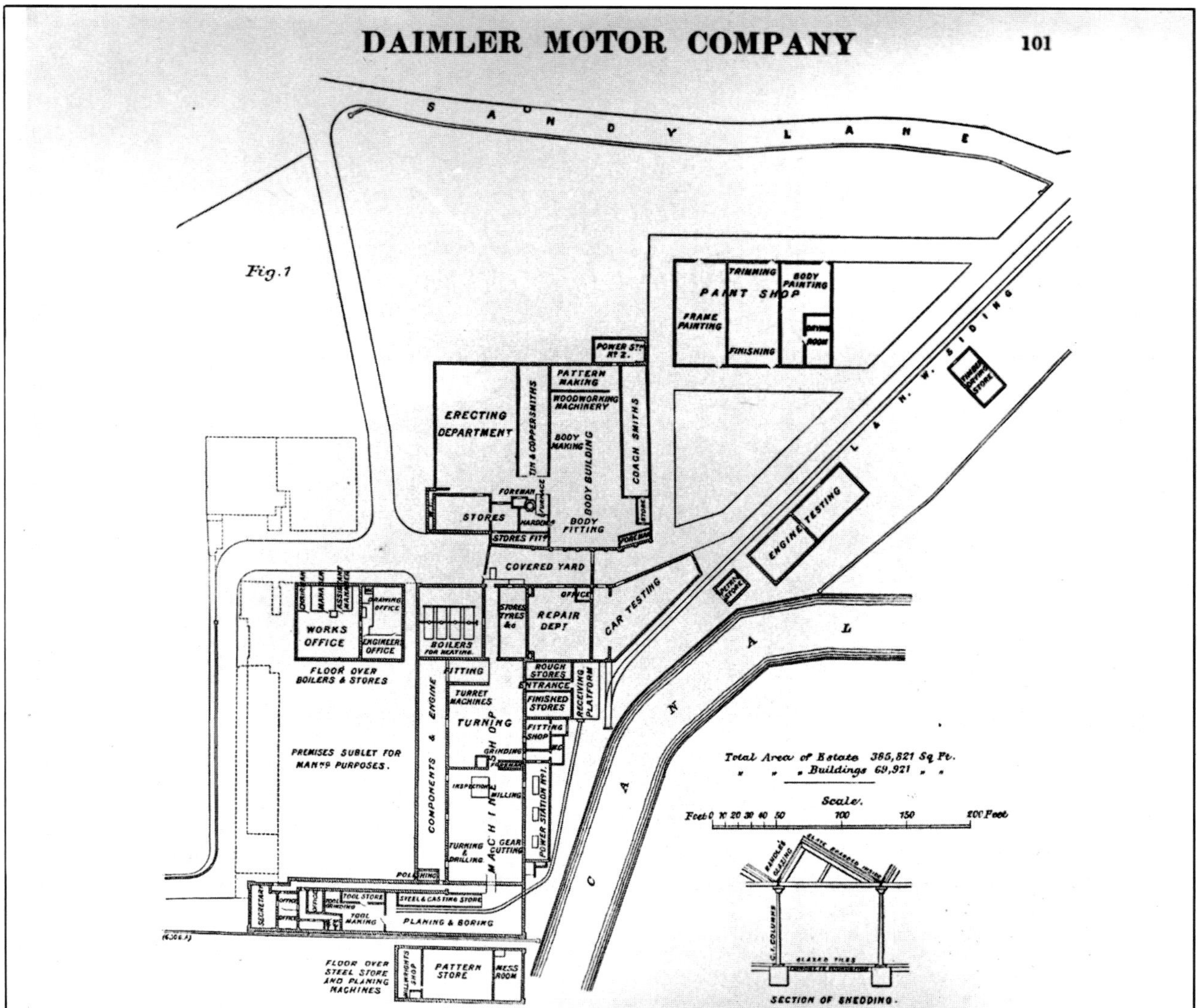

Plan of the Daimler Motor Company's Works, Coventry

Finchley garage run by Alfred Burgess - MMC's Manager and former Company Secretary.

The end of the MMC should perhaps be regarded as a sad one, and although originally formed by Lawson, not necessarily an inevitable one. The GHCC affair was something of a joke, but the MMC was a far more serious proposition and ran in a more thorough and businesslike manner. In its short but eventful life, it is thought that over 4,000 vehicles were sold. Only George Iden seems to have made anything from the experience, however, as in 1904 he established the Iden marque, first in Coventry's Parkside, and later in Fleet Street. Production of both cars and commercials ended during the 1907 season.

At the first AGM of the Daimler Motor Co. (1904) Ltd, held in November 1905, Sir Edward Jenkinson delivered a very different report from those of the previous meetings. For the 1904-05 trading season, profits had risen to £83,000 - nearly £76,000 up on the 1903-04 figures.

King Edward VII ordered another Daimler during 1905, his seventh so far. The Prince of Wales (later King George V) also bought a 28/36hp Daimler to add to the two cars he had bought in 1903.

More Sporting Achievements

Various victories came once again in competition, noteably in the speed trials held at Bexhill, Brighton, and Blackpool. Hillclimbing was another favourite pastime with Daimler owners, Ernest Instone and Percy Martin always gaining good results and a large number of trophies.

It is interesting to note that during 1905 a pair of cars were built by the Company to compete in the first ever Tourist Trophy race, held on the Isle of Man in September 1905, although they were never actually used for this purpose. They were the first propshaft-driven models produced by Daimler since this form of transmission had proved so unsatisfactory in the 1901 trials, and the radiator was also a notable feature.

For some years, horizontal fins had been running through the Daimler radiators to aid cooling, and although this arrangement was still used on some of the 1904 models, it was in this year that the traditional

Top of Page:
Mr Neville aboard one of the two Daimlers built to race for the Tourist Trophy event in 1905.

Right:
A 1904 28hp Daimler.

Right:
Mrs Maud Manville pictured in August 1905, driving her 35hp Daimler, DU 541.

fluted grille appeared - a feature still found on Daimler cars to this day. Eventually, one of the TT cars was sold off to a Mr H.H. Neville after four years sitting in the Experimental Department, but what happened to the sister car, however, is still very much a mystery.

A Reduction in Prices

During 1906, there was a considerable reduction of about 15% in the price of Daimler cars. In a Press Release, the Company gave the reason for this as being *"entirely due to the great increase in output, economy and production, that the Company is able to announce material reductions in the prices of its carriage."*

In fact, virtually every manufacturer in the country had brought its prices down. However, this attempt at making the cars more competitive, reduced the chassis price of the 8.5 foot wheelbase 28hp model from £700 to £590; the 30hp chassis from £825 to £690, and the 35hp chassis from £1,050 to £890. This helped Daimler to build and sell around 500 cars in 1906, making them the UK's largest producer at the time.

Instead of appointing agencies throughout the country, it was decided to open branch depots at some of the larger cities - Nottingham and Manchester being the first two to be selected. Markets were also being cultivated on the Continent, and agencies were established in various European cities. The Daimler was, for a short time, officially introduced into America as well.

Below:
The 45hp Daimler at the Museum of British Road Transport in Coventry with the DU 541 number plate. This, however, is not thought to be the car originally bearing that number.

The Retirement of Sir Edward

It was during 1906, that the Company suffered a great loss - the resignation of the Chairman and Daimler's guiding light, Sir Edward Jenkinson. The improvement in the fortunes of the Company can largely be credited to this one man, although his reasons for retirement are still surrounded in mystery. It has been said that it came about when a well-meaning Daimler official suggested that he would do well to buy some more

Above:
Ernest Instone won the first ever hillclimbing event at Shelsley Walsh for Daimler in 1905. Here he is racing a 35hp Daimler again at the start of the Sunrising Hill course.

Below:
A 38hp landaulette by Windovers, ordered by Viscount Hirchingbrooke, Earl of Sandwich, in 1906.

shares in the Company - Sir Edward was apparently offended by the very thought.

Officially, he gave his reason as *"finding the work too exacting now that the Company was expanding"*, but he was a very capable businessman. After a failed attempt to get Sir Edward to reconsider (he resigned from the Board as well) the Company gave him a cheque for 1,000 guineas, and a specially commissioned portrait in recognition of his services. The important position left by Sir Edward was filled by Edward Manville (later Sir Edward Manville JP, MP), who would hold this office with similar distinction for a great many years.

The 45hp Daimler

In 1906, a firm called the Société Anonima Officine de Luca Daimler was started up in Italy, most of the share capital being held by Daimler in Coventry. At the same time a new bonus scheme at Coventry proved extremely popular with the workers. A non-profit making course to train young men in the field of automotive engineering was also introduced, and many prominent motoring figures passed through the scheme.

The Gate Speed Control designed by E.W. Lewis and patented by Daimler was not only used by the Company themselves, but Rolls-Royce, Swift, and the Climax Motor Co. - a Coventry car manufacturer which built machines from 1905 to 1907.

Above: *Barker bodied Shooting Brake built for Mr W. Campbell.*

Other outside business included an enquiry from Saunders regarding racing power boat engines.

A new and very attractive 45hp model (the Board had asked for a 50hp car) was produced during the summer of 1906, some of its details being quite unlike those of the more usual Daimler practices. Until now, Daimlers had always employed frames of flitch-plate, channel steel and wood, but for this vehicle, a pressed steel frame was used, curved slightly upwards over the rear axle line.

All four cylinders were cast in one, and both the bore and stroke were 150mm. It also had a ball bearing crankshaft which, at the time, was a very rare feature in such a large engine. Although both the inlet and exhaust valves were mechanically-operated, the exposed camshaft was retained for ease of maintenance. With a wheelbase of nearly 10 ft it was a particularly handsome and serviceable car.

The *Autocar* noted: *"The general impression conveyed by an examination of this vehicle is that so far it is the finest vehicle that has ever been turned out by the Daimler Works, and it will bear comparison with the*

Below: *This Barker bodied limousine was built for the Duke of Richmond and Gordon, circa 1907.*

Left:
Lord Howard de Walden and "Daimler II", photographed during the 1907 La Perla del Mediterraneo. Daimler engined boats took several titles in this period, including the Dewar Cup and the May Cup.

Right:
Ernest Instone on his 35hp Daimler after winning the Gottlieb Daimler Memorial Trophy at the first ever Brooklands race meeting on the 6th of July 1907.

Left:
One of the three cars entered by Daimler for the 1907 Targa Florio, this one is driven by Hemery. It is not widely known that one of the Daimler staff was killed during practice.

Right:
A Daimler at the start of the 1907 Kaiserpreis Race.

Below:
Brooklands 1908. Although by this time the Company had decided not to continue with their racing exploits the cars were still very competitive in the hands of enthusiastic amateurs.

latest and most advanced designs. We anticipate that it will render a good account of itself in coming competitions."

The Range for 1907

The 10.6 litre 45hp car was joined by updated versions of the 28hp (120 x 150mm), 30hp (130 x 150mm), and 35hp (140 x 150mm) models. In their article on the new models, *Autocar* stated:

"No departures have been made in general principles, improvements being in construction and modified design, and, this being so, it only remains for us to deal with the alterations and modifications which have been effected to produce a mechanical road vehicle which but a few short years ago would have been regarded as a step beyond the ideal."

The article went on to give a glowing report of the Daimler models, and said that he had no doubt that they would be a great success.

Type	*bhp @ 1000rpm*	*Chassis TL*	*Chassis TM*	*Chassis TO*
28hp	*43*	*£550*	*£600*	*£650*
30hp	*55*	*£620*	*£670*	*£720*
35hp	*65*	*£720*	*£770*	*£820*
45hp	*74*	*£820*	*£870*	*N/A*

From a collection of contemporary postcards of the Daimler works during the period 1907 to 1908.

Top: *The Brass Machining Shop*

Centre: *The Chassis Erection Shop.*

Bottom: *The Final Assembly Shop.*

Above: *The Despatch Shop.*

Below: *A Daimler racing car, presumably prior to leaving the works.*

Left:
The Renard Road Train.

Below:
A 1909 private car chasssis converted to a fire engine by the Aldershot Volunteer Fire Brigade.

A third branch depot was opened, this one at Bristol, and for the first time, British built Daimlers were exhibited at the Paris automobile show, as well as at the Berlin exhibition. Enquiries were coming in from such distant parts of the world as South America, the Straits Settlements, and the Far East. Indeed, the King of Siam took delivery of a 28hp Daimler during March 1906, and although the body was of the long since obsolete Roi-des-Belges type, it was of a somewhat revolutionary design at the time. With coachwork by the Lacre Motor Car Company of Poland Street, London, it was designed to seat seven; two in the front, three on the back seat and two in the tonneau, for whom two revolving chairs were provided. Another feature was the canopy which incorporated roll-up side screens.

Once again, Daimler came into conflict with the British Motor Traction Co. in the Courts. It will

Above: *King George V was loaned this 57hp limousine (chassis TD6314) by the Daimler Company during the early part of 1909. After touring Cornwall in it the the King decided to buy the car.*

Left:
A 1908 "Harting" landaulette very similar to the 42hp car supplied to the Queen of Spain for Her Majesty's use while staying in England.

be remembered that they had registered the Daimler Wagon Co. Limited in 1901, and evidently encouraged by their success at fighting off an injunction on that occasion, tried to do the same thing again by registering a firm called the London Daimler Company with a capital of just seven £1 shares!

With more money at their disposal, it was decided to fight to protect the now highly regarded Daimler name, and an injunction was sought in the High Court. The fact that Milnes-Daimler, Mercedes-Daimler (Daimler-Benz) and of course, the Daimler Motoren Gesellschaft were still trading, left the Court in no doubt that they should refuse the injunction on the grounds that the Daimler name was already in use elsewhere. However, not satisfied, the case went to the Court of Appeal, and in April 1907 the decision was reversed, although the Coventry company had to pay all of the costs due to the London outfit having no money.

Above: *Charles Yale Knight aboard a sleeve valve 15hp Daimler in Coventry's Sandy Lane.*

A Move into Commercial Vehicles

The Board had made a decision some time ago that they should move into commercial vehicles as soon as possible. As early as December 1905, the Board had authorised a joint-project to build a bus with the United Electric Car Company. A licence for the manufacture of the Auto Mixte system of petrol-electric transmission for heavy vehicles was obtained, and very soon after, several experimental machines were up and running. These prototypes proved to be more than satisfactory, and the Daimler works was extended to allow for the extra production capacity needed. A large plot of freehold land was also purchased near to the existing Daimler site (on the Radford estate) for it was felt that even if the commercial vehicle venture failed, the land would soon be needed anyway for the rapidly growing car business.

1906 had been a very good year, and although 1907 witnessed a slight decrease in profits, this also proved to be another period of prosperity. At the AGM, held for the first time at the Coventry works, it was announced that a new service depot at Highgate would replace the old Brownlow Mews one, and that the Company now employed something like 2,750 people.

The Company also continued to be honoured with their Royal patronage. King Edward VII ordered two more cars in 1907, taking his total to nine Daimlers. On this occasion, he bought a Shooting Brake by Hooper, and a 35hp landaulette. The Duke of Connaught ordered two chassis, and whilst visiting this country, the King of Spain also bought a new Daimler. Business had increased so much in Spain in fact, that the Company opened service depots in Madrid and San Sebastian - orders were coming, quite literally, from all over the world.

Competition News

In competition, the Daimler marque did so well that it led to allegations of the Company using professional drivers instead of private entrants - in 1907 alone, Daimlers took Class wins in 23 recorded events. Daimler cars were a regular sight at Brooklands for many years (Ernest Instone winning in the first race meeting held there in July 1907), and Daimlers were still winning at the Weybridge track in 1914 with Oscar Morris and his 8litre Daimler powered Darracq, once the property of Sir Algernon Guinness. Even in the United States, a 45hp Daimler had taken victory at Giant's Despair in the previous year, a steep climb renowned for its twists and turns.

During the 1906-07 period, they were seen in several important events, including the Herkomer Trophy (at least nine Daimler cars had been entered), the Graphic Trophy, the Kaiser Cup race, and the gruelling Targa Florio.

The Kaiser Cup had not been good for the Company, the three drivers (Ison, Bush and Hodierne) failing to make any real impression, despite their specially-designed engines. Only Ison made it through to the finals, but was forced to retire with the clutch problem that had dogged the Daimler team. The Targa Florio, however, was an excellent event for Daimler, moreover it was a true test of the cars ability.

Below: *How a sleeve valve engine works, as described in a September 1908 issue of the Autocar.*

THE AUTOCAR. SEPTEMBER 19TH, 1908. 431

running of the new engine could be equalled by that of the ordinary type it might still surpass it for the simple reason that its distribution gear should be practically wear-proof. Hence the main bearings and big ends, instead of being the last thing to require attention, should be the first, as the valves and valve gearing, owing to the very light work they are called upon to do and the almost ideal conditions under which they work, should run almost indefinitely without the necessity for cleaning, adjustment, or renewal.

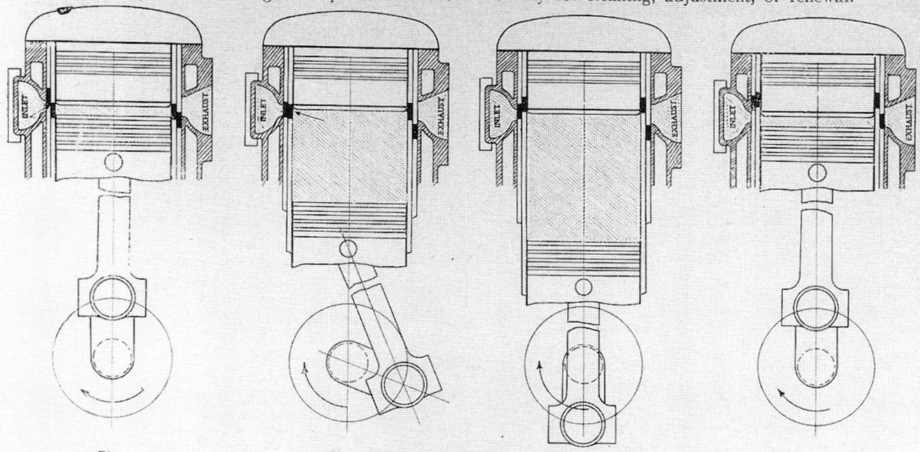

Fig. 1. Fig. 2. Fig. 3. Fig. 4.

Fig. 1.—Commencement of the suction stroke. The outer sleeve is travelling downwards and the inner one upwards, making a rapid opening of inlet.

Fig. 2.—Maximum opening of the intake. The crank pin has described an arc of 105°. The port opening at this point is 12 x 140 mm.

Fig. 3.—The piston is now at the bottom of the stroke; both sleeves are travelling upwards. The intake port is still about 3 mm. open to permit the continued entrance of gas into the cylinder through its own inertia. The ports then close immediately.

Fig. 4.—The piston is at the top of the stroke; both the inlet and exhaust ports in the inner sleeve are over the broad ring in the cylinder head, effectually making a compression tight chamber.

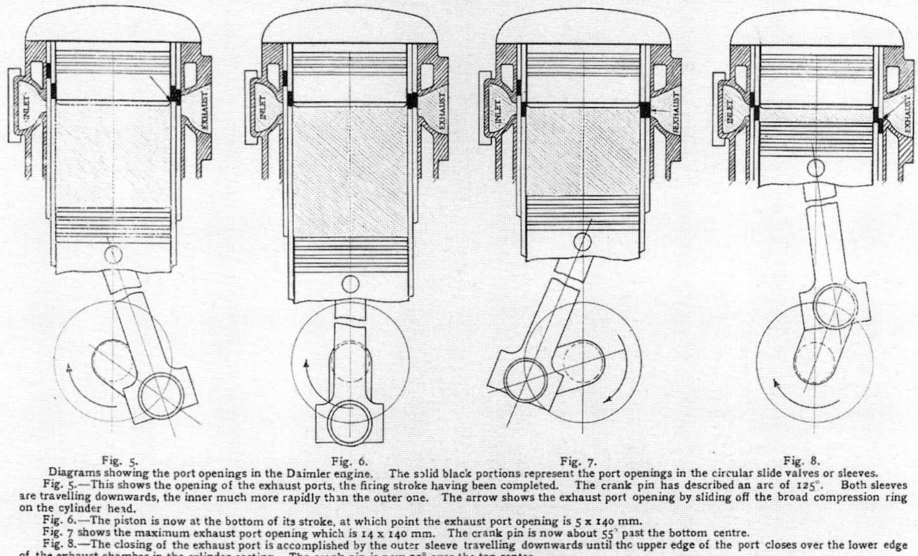

Fig. 5. Fig. 6. Fig. 7. Fig. 8.

Diagrams showing the port openings in the Daimler engine. The solid black portions represent the port openings in the circular slide valves or sleeves.

Fig. 5.—This shows the opening of the exhaust ports, the firing stroke having been completed. The crank pin has described an arc of 125°. Both sleeves are travelling downwards, the inner much more rapidly than the outer one. The arrow shows the exhaust port opening by sliding off the broad compression ring on the cylinder head.

Fig. 6.—The piston is now at the bottom of its stroke, at which point the exhaust port opening is 5 x 140 mm.

Fig. 7 shows the maximum exhaust port opening which is 14 x 140 mm. The crank pin is now about 55° past the bottom centre.

Fig. 8.—The closing of the exhaust port is accomplished by the outer sleeve travelling downwards until the upper edge of the port closes over the lower edge of the exhaust chamber in the cylinder casting. The crank pin is now 20° over the top centre.

The arrows in figs. 2, 5, and 7 are to draw attention to the positions of the sleeve ports, and not to indicate the flow of the gases, which in each case are in the opposite direction to that pointed by the arrow.

On the 24th of August last a party of three Russian automobilists, all members of the St. Petersburg A.C., started on a tour of Continental Europe with the object of interesting their fellow countrymen in motor touring, and to prove the possibility of making an extended journey by motor car. The itinerary is St. Petersburg, Riga, Berlin, Brussels, Paris, Lyon, Nice, Genoa, Vienna, Riga, and St. Petersburg.

In 1906, only ten cars had entered the Targa Florio, but 1907 attracted 46 competitors, which included three Coventry-built, eight-litre four-cylinder Daimlers. These were not the 45hp models that one might expect, but 30hp models supplied to De Luca at £400 each. One of the Daimler staff, a Mr Rueger, was killed during testing, but morale was raised when George Ison (the Works Tester who would be with the Company for over 25 years) held fourth place for a while. Eventually, he finished 13th, whilst Le Blon was 20th, and Hemery 23rd out of thirty finishers.

The Company were going to enter in the Heavy Section of the Tourist Trophy Race for 1907, but cancelled their plans at the last minute. At least in 1906, a decision had been made not to compete from an early stage in the proceedings.

In the meantime, Daimler powered boats were also spreading the Company's fame, with several notable results to their credit. In October 1906, it was *"resolved to supply Lord Howard de Walden with three special six-cylinder 50 hp motors for a Motor Racing Boat at an inclusive price of £1,000."* These were used to propel "Daimler I", the first of his racing boats built to compete with largely Napier, Mercedes, Fiat, and Panhard & Levassor powered machines.

The 1907 Perla del Mediterraneo was an event ran alongside the Targa Florio for power boats. Lord Howard de Walden (it is interesting to note that the Hon. Evelyn Ellis was de Walden's son) had entered two Saunders boats, Daimler II and Daimler III, each fitted with three 57hp engines. No success came in this particular race, but the Dewar Cup and the May Cup came de Walden's way later in the year. Daimler II, incidentally, had the distinction of being entered in the 1908 Olympic Games, but sadly it hit problems, missing out on a certain medal chance.

Links with Renard

With the future in mind, commercial vehicles continued to play an important part of the Daimler business and in 1907, an agreement was completed to enable Daimler to manufacture the Renard road-train. Until 1907, the production of commercial vehicles had been very low, with the only significant examples being a few two and three tonners and an 18hp fire engine.

The Renard Road Train, however, gave Daimler entry into the commercial field with a proven, if somewhat unrefined, product. As the name "Road-train" suggests, the vehicle was a train type, with a power unit up front which transmitted drive to each of the trailers via a series of propshafts. There were usually between four and six trailers attached to each tractor unit, and the front wheels of each were arranged in such a way that they followed the track of the preceding one. It was essentially a vehicle for overseas use, as the British roads were not really capable of accommodating such large machines. After a certain amount of development by the Company, several units were sold to customers in Europe, North and South America, Australia, Spain, and even Egypt.

Car Illustrated noted during August 1910 following a factory demonstration: *"A great many of us who were present had been acquainted with the old system of traction, as applied on the Renard principle, and therefore were in a position to appreciate the vast improvements which have been effected by the Daimler Company. Perhaps the most astonishing feature of all, when watching these huge motor trains moving on smooth road or rough field, is the wonderful silence of the whole mechanism..."* As regards the engine, it was *"of the well-known Daimler Knight type, having six cylinders cast in pairs and listed as 80hp. There is nothing in this engine that materially differs from the 57hp pattern except the cooling system"* which was made more efficient for work in warmer climates.

Another type of motor vehicle designed and manufactured at around the same time was a rail coach powered by a petrol engine. Several were sold to the old London, Brighton & South Coast Railway Co. and for a while, they were in considerable demand. Later, in 1908, a 34 seater bonneted double decker bus was constructed using a 30hp four cylinder petrol engine for motive power, Auto-Mixte petrol-electric transmission and a worm driven back axle. This was to be the only exampleever built, although it did at least serve as a good prototype for the KPL petrol-electric bus that was to follow.

The Sleeve Valve Engine

The six-cylinder engine was first introduced by an established motor car designer, so too were the Wilson gearbox and the fluid flywheel. The sleeve valve engine, however, and the circumstances behind its introduction, were completely different. Its inventor, Charles Yale Knight, despite his flair for engineering, was untrained, unqualified and comparatively inexperienced. In fact, he knew little more than any average car enthusiast.

Nevertheless, this amateur (for he was little more) was responsible for an invention that would rock the entire motor industry. For months, the

Above: *The Silent Knight 15hp Daimler. Announced in November 1909 it was without doubt the Company's most important model during this period.*

Below: *Joseph Cockshoots of Manchester were responsible for the body on this delightful 15hp model dating from 1910.*

Above:
The Daimler bodyshop circa 1910. Note the way in which the body lifts to provide maintenance access.

sleeve valve engine was the topic of debate amongst some of the country's most eminent designers. While some regarded it as a significant advance over the poppet valve, there were others who saw it as nothing more than an added complication which would cause lubrication difficulties and increase manufacturing costs.

While discussion raged on, the subject was beinglooked at in detail by several French, German, and American firms, but the first to appreciate that the sleeve valve represented an important step forward from the widely used poppet valve engine was the Daimler Company.

Knight was a young American who had spent most of his early years in Northern Wisconsin. As a boy, he was employed in a local saw mill, but he soon realised that his future lay in agriculture. He decided that his mission in life was to highlight the many hardships experienced by the farmer, and opened his campaign by contributing articles on the subject to the leading newspapers in the States. This introduction to journalism led him to a job in Washington, and it was here, in 1901, that he first became interested in the automobile.

It seemed to Knight that the biggest problem with the petrol engine was the awful noise it made. Designers at this time were more interested in getting their units to run, than run quietly. The weakness was the cam striking a valve tappet, and so Knight set himself the task of designing a completely new type of valve gear which didn't require a tappet and cam to make it work.

Like Lanchester before him, Knight studied everyrelevant technical work, not for ideas, but to steer clear of beaten paths, and avoid the many pitfalls into which others had fallen. Eventually, he came into contact with an experienced mechanic by the name of L.B. Kilbourne who had been in the trade for some 15 years. The two men carried out many experiments, and registered a patent in 1903 (No. 14729) for their invention on a new type of valve gear. However, it was soon realised that it would not be suitable for a high speed engine, and would almost certainly make even more noise than those already available.

Then the idea came to him - why not a cylinder within a cylinder? Instead of trying to move the heavy walls of the water jacket, why not employ cast iron liners the full length of the cylinder, and move them up and down over the inverted piston, using a wide ring on the bottom of the latter to cover the ports at the time of compression and ignition?

Experimental work began on this new engine during the August of 1903. After trying quite successfully with just one sleeve, the gear to move it was found to be too complicated to be quiet. A second sleeve was introduced, and after a whole winter of frustration, a four cylinder engine of 3.5inch x

4inch bore and stroke, was put to the test.

Completed in October 1904 it was fitted in a Panhard, which the partners drove for many thousands of miles without any real trouble. Thus encouraged, they decided to go into manufacturing complete vehicles, with the first being ready for sale during the spring of 1905 - known as the "Silent Knight", it was fitted with a 40hp sleeve valve engine.

Knight's introduction to the Daimler Co. came about in a rather unusual way. Early in 1906, a friend of his happened to meet Sir Edward Manville in London. It was mentioned that this extremely quiet new engine had been produced, and Manville requested that some paperwork be sent to the Company. All were impressed by this bold design, and Knight was instructed to bring one of his cars to Coventry for testing.

The friend stressed to Knight the importance of the request, and the fact that a major firm like Daimler would never make such a suggestion unless they were genuinely interested. With business being slow in the States, the invitation was accepted. Soon, Knight was in Liverpool, and from there he drove down to the Coventry works. Percy Martin was the first to examine the engine and test drive the car. The Company was favourably impressed with the design, especially its quietness, but it was clear that the construction was, at this stage, far from perfect.

Eventually, after much work, the engine was made into something quite exceptional, and Daimler negotiated the exclusive rights for manufacture in the UK, as well as a licence to export to all corners of the globe. The old poppet valve power units were abandoned almost immediately in favour of the Knight engine, but with this news came problems.

As with most new things in a world frightened change, it was bitterly attacked, mostly, as is often the case, from people who had no experience whatsoever of the Knight engine. As lubrication seemed to be the biggest problem in most minds, and the fact that great damage would be inevitable in the event of it running low on oil or water, lectures were held to try and convince pundits of the engines merits. One of the most interesting was given by Charles Knight himself during October 1908, to members of the RAC.

RAC Tests on the Knight Unit

Although all this controversy drew a lot of attention to the Knight engine, it could hardly be regarded as good publicity. A strong defence had been set up in some quarters, but it was felt that such strong prejudice against the engine could be damaging to the Company if allowed to continue. It was decided therefore, that a test should be devised, so severe that it would silence all of the critics.

At one of Knight's lectures, a test of 10,000 miles was suggested by a Daimler owner, but the reply came *"the public will soon have a test before them beside which a 10,000 mile road test would be only a drop in the bucket."*

Two engines were selected, one 38hp and one 22hp version, respectively having a bore and stroke of 124mm x 130mm and 96mm x 130mm. For the first part, the test consisted of running the two Knight engines continuously for no less than 133 hours whilst under the close observation of the RAC. They were then removed from the test bench and installed in two Daimler cars, complete with five seater bodies, which were then driven under their own power from Coventry to Brooklands for a further test of some 1,920 miles. After this, the cars were taken back to Coventry and the engines taken out, where once again, they were subjected to another bench test, this time though for five hours of brake horsepower testing. Finally, they were dismantled so that their condition could be examined after the gruelling trial, and a full report could be submitted by the RAC engineers. The test began on 22nd March 1909 at six o'clock in the morning, and after several days and the equivalent of countless thousands of miles, the results were outstanding.

This extraordinary performance seems even more remarkable when one considers the relatively primitive metals available, and it is doubtful whether any poppet valve engine of that time could have stood up to such a harsh test. The Daimler Motor Company therefore challenged any manufacturer to try their engines against the fine results put up by the sleeve-valve units - the invitation was left open for three months, but not one company took up the gauntlet.

As soon as the RAC certificates were made known to the public, the criticism of the Knight engine quickly died down. Ironically, the Daimler Motoren Gesellschaft purchased a licence to manufacture under the Knight patents, as did Minerva of Belgium. Panhard & Levassor were ardent supporters of the sleeve valve, and were the last manufacturer to revert back to poppet valves, not doing so until after the Second World War. Suddenly the sleeve valve had come into fashion. Belsize of Manchester even produced their own sleeve valve system, using a single sleeve that turned in a semi-rotary motion - it was claimed to be even better than a Daimler engine.

Whilst all of these important events were taking place, the Company was not standing still. With the erection of several new buildings and numerous problems with the workers' trade unions, 1908 saw a slight loss, but during the following year, this was converted into profit, thanks largely to the development of foreign markets. It is interesting to note that Daimler had been exporting to Japan before that country's industry had even been established, and that inroads were being made into the American market.

The Daimler Company of New York was established, and the Company was still exhibiting at the New York Show in the years running up to the First World War. It seems strange that the Daimler marque would become virtually unheard of in the U.S.A. in modern times.

Despite the high duty rates for British manufacturers exporting to the Continent, sales were increasing rapidly, prompting the Company to form a French concern which eventually became known as the Société Francaise de la Daimler Motor Company, based in Paris.

Originally established in this "selling Company" underwent many changes before adopting its final title. Léon Girardot, the racing driver who had helped found the CGV marque in 1901 (which became Charron from 1907), was employed as the firm's representative. Girardot had been in close contact with the Company for a number of years, Daimler having given him a car in which to attend the 1908 French Grand Prix.

De Luca Daimlers in England

Daimlers had also been sold in Italy for some time, with the models being christened "De Luca" Daimlers. The Company in Italy had been founded in 1906, with the finance largely coming from Coventry.

It seems as if Daimler in England developed the 17/21hp engine for De Luca, as this was their most popular model, and yet they were sold in the UK *"with discretion"* as Percy Martin put it. At least the 25hp chassis of the 17/21hp type were built in Coventry, and other cars were produced on their behalf as well, such as the Targa Florio machines. During May 1908, two of these were imported into the United Kingdom for the first time, and entered by the Daimler Motor Company for the 2000 Miles Trial. Basically the same as their British counterparts, save for a few fairly minor details, these 17/21hp machines quickly found favour with the motoring press, especially in view of their simple operation.

Cars purchased from Italy were sold strictly as De Lucas from 1908, with any reference to Daimler being deleted on the imported cars - they even had their own stand at Olympia in 1908. Over the next couple of years, production of the Italian cars was gradually phased out, although there had been quite an extensive range available, right up to a massive 58hp model.

Types for 1909

There were six standard Daimler models for the 1909 season. The first column relates to the chassis designation (length and horsepower rating), the fourth column shows the wheelbase, and the final column is the cost of a basic chassis. Extra was charged for magneto ignition and wire wheels - the rest is self explanatory.

TB.22	*4 cylinder*	*Live-axle*	*9.5'*	*£425*
TC.22	*4 cylinder*	*Live-axle*	*10.5'*	*£475*
TB.38	*4 cylinder*	*Live-axle*	*9.5'*	*£525*
TC.38	*4 cylinder*	*Live-axle*	*10.5'*	*£575*
TC.48	*4 cylinder*	*Chain-drive*	*10.5'*	*£675*
TC.57	*6 cylinder*	*Live-axle*	*10.5'*	*£765*

More new models

During the early part of 1909, much attention was given to the new 15hp model, fitted with the Silent Knight engine of 78mm x 100mm. It was a smaller car than had been built by the Company for some time, and it was hoped that it would be ready for the forthcoming Motor Show to be held in November. The standard touring body was very attractive, and it could be supplied with three different gear ratio options to suit the body weight. The 15hp cars were as follows:

Model A (Roadster)	*Phaeton*	*2str*	*£425*
Model B (Roadster)	*Phaeton*	*4str*	*£435*
Model C (Touring)	*Phaeton*	*4str*	*£445*
Model D (Touring)	*Phaeton*	*6str*	*£455*
Model E (Touring)	*Limousine*	*4/6-str*	*£500*
Model F (Touring)	*Limousine*	*4/6-str*	*£525*

Other models for 1910 included a 3.76litre 22hp, a 6.28litre 38hp, and the Company's first vehicle to employ a six cylinder engine, the 9,421cc (124mm

Left:
A flush sided Daimler tourer from around 1910 in suitably majestic surroundings.

Below: *One of the Daimler cars taking part in the "Herkomer Trophy" of 1907.*

x 130mm) 57hp, introduced with poppet valves.

Only a couple of years earlier, Daimler had strongly denied rumours that they were building a car of this type, but they had obviously been true. Perhaps the Company wanted to keep the project quiet due to the fact that they were having great difficulty in making the new car run properly. Indeed, the first dozen cars built were found to have such bad torsional vibration, that they could not even consider selling them. However, a new feature on all of these was the use of only three forward speeds (as on the 15hp model). Although a couple of years later, this was changed back to four speeds.

The prospects of the Company were thoroughly sound. The definite and individually specified orders, up to the end of November 1909, exceeded the total sales for the whole of 1908. In addition, and with more than a little irony, Dr F.W. Lanchester was appointed the Company's Consulting Engineer during 1909, one of his first jobs being the introduction of a crankshaft vibration damper on the new six cylinder car. At a later stage, when the coming together of the Daimler and Lanchester marques is described, the incredible history of this great man will be described. In the meantime, however, there was another amalgamation just around the corner.

CHAPTER FOUR

THE AMALGAMATION WITH BSA

Although the actual birth of the Birmingham Small Arms Co. Limited was in 1861, its origins can be traced back a further 150 years. Until 1692, virtually all of this country's military arms were manufactured in Holland. But for a remark made by the King, stating that he regretted England not producing its own weapons, the situation might have remained this way for quite some time after.

It was thanks to Sir Richard Newdegate, the MP for Warwickshire, that an order was placed by the Government with five of Birmingham's leading gunsmiths, for 200 muskets a month. This continued almost uninterrupted for over 150 years, until some 14 of the master gunsmiths engaged on Government work, formed an association known as the Birmingham Small Arms Trade, to try to compete with the Government run factory at Enfield. It was from this group that the Birmingham Small Arms Company was founded in 1861.

For the first two decades of its existence, the Company dealt solely with orders for military arms, not only for this country, but for foreign governments as well.

After a brief (and largely unsuccessful) entry into building their own vehicles in 1907, the Directors felt that efforts should be made to gain a controlling interest in a car manufacturing business, whose products were on a par with those of their own firm. At that time, the choice was not a particularly wide one, but eventually negotiations began with the Daimler Motor Company with a view to an amalgamation of interests.

Facing page: *A 1914 model Daimler 'Frankford' Landaulette on the 20hp chassis, pictured here outside the Fords Hospital in Coventry.*

Above: *A 1909 BSA 25/33hp landaulette.*

Right: *This 1909 TA 23 now forms part of the Jaguar-Daimler Heritage Trust.*

A huge plot of land had been purchased in the Small Heath area of Birmingham during 1862, in order to build a new factory when it was assumed that the Government would place a large order as soon as the Company could deal with it. Unfortunately, they waited for something like five years for their first contract - the conversion of 100,000 muzzle-loading Enfield rifles. Two more years were to pass before an order was placed for arms actually manufactured in the new works.

In February 1873, the Birmingham Small Arms & Metal Co. Limited was registered, and a new factory at Adderley Park was acquired. The Company continued to trade under this name until 1896, but during this year the munitions part of the business, and the parent company reverted back to its original title.

Business continued at a moderate rate for a year or two but during 1879, work began to slow down at an alarming rate, leading to the closure of the Small Heath factory. There was little need for rifles or other machines of war as, for the first time in recent history, the country was at last at peace. Machines previously used for turning out weapons by the score were standing idle, and the BSA was forced to seek other fields of activity.

By the end of 1880, BSA had taken up the manufacture of good quality bicycles, and it was while the Company was thus engaged that a curious coincidence occurred which seemed to foretell the eventual coming together of the BSA and Daimler concerns. In 1881, BSA made a number of cycles, many of which were displayed at the old Stanley Show.

Above: *A Royal Daimler. This 57hp limousine by Hoopers was delivered to HM the King in September 1910.*

Below: *Dudley Docker circa 1916.*

Among these was a tricycle designed by Martin Rucker - one of the three businessmen who acquired the original Daimler patents from the Daimler Motor Syndicate in 1895. The other coincidence involved none other than Harry Lawson, for in the June of 1884, he submitted his design proposal for a so-called "safety cycle" (his first "safety" had been designed in 1876) but rather oddly, BSA rejected it.

For the next decade, the popularity of the bicycle developed, with BSA devoting more and more effort towards cycle production, and the manufacture of various related components. A new form of rifle was introduced in 1888, however, and during the following few years, the whole of the British Army was re-armed with this weapon. For a little while at least, the demands on BSA became so great, that it had to abandon its cycle trade.

From 1893 to 1908, the Birmingham Small Arms Co. confined its cycle business exclusively to manufacturing parts for the industry. In 1894, the works were enlarged, and BSA quickly acquired a good reputation for quality; the trademark of three piled rifles was to become a symbol of superb workmanship. During 1908, after an interval of some twenty-one

Above: *One of the early sleeve valve cars, this 1910, 12hp tourer with its Knight 1.7 litre, four cylinder engine is still in superb running order today and has formed the basis of a Matchbox diecast model.*

Below: *A "Cloverly" doctors coupé body on a four cylinder Daimler chassis, 1910.*

years, the Company once again introduced a complete BSA bicycle, with the first BSA motorcycle following just one year later.

Meanwhile, the motor car industry was by now well established. However, the BSA Company, despite periodic rumours that it was intending to supply car manufacturers with components, remained outside this field of activity. The problem was that the design of motor cars varied to a far greater degree than bicycles, so it was not possible to indulge in the mass production of general components.

The situation at Daimler

During 1910, Daimler had an extremely successful year, with profits soaring into six figures and a new 12hp model that looked set to be a big success. Although its fortunes were still inclined to fluctuate somewhat, Daimler was without doubt the leading motor manufacturer in this country.

By this time, all of the Daimler cars employed sleeve valve engines (including the big 57hp model which had originally used poppet valves), and the Lanchester-type worm-driven rear axle. Another six cylinder car was added to the range - the 5.6litre 33hp, although it did not appear in the 1911 range. Royalty once again had honoured the Company.

King George V took delivery of two superb Hooper bodied 57hp limousines, and Queen Mary ordered a four cylinder 38hp limousine. A 57hp landaulette, which had been ordered by King Edward VII before his death, was eventually completed by Hoopers, and delivered to Queen Alexandra.

Also during 1910, Daimler had embarked on an ambitious programme of the full scale production of

Below: *Daimler diversity. The revolutionary KPL bus, loaded to bursting point with local hooligans!*

Left:
The demand for heavy tractors soon became widespread, and in 1911, Daimler designed and built a number of these in conjunction with Fosters of Lincoln for agricultural and other use.

petrol-electric buses for public service. These were built under the patents of Messrs Knight, Lanchester, and Pieper. Pieper was a Belgian who had made his name in the very early days of the motor car by designing a small two seater driven by a De Dion engine; it had a very long belt drive, and was rather attractive for its day.

However, the buses, known simply as KPLs, had a number of important features, some of which were well ahead of their time - perhaps the most interesting concerned the unit construction, executed in steel. All of the seats faced in the direction of travel, which was something of an innovation in those days, and the driving system had two distinct power-units, consisting of petrol engines and electric dynamotors.

The separate 12hp Knight-Daimler engines and dynamotors coupled direct, were slung from brackets on either side of the bus. The transmission was then continued on either side by means of two universally-jointed shafts and worm-gearing to the two short live axles carrying the rear wheels. One result of this system was that no differential was required, the electric balancing taking its place.

A magnetic clutch and brake formed part of the

Below: *One of the tractors at work in Warwickshire.*

Above: *This incredible machine, a lengthened 38hp six cylinder Daimler, was bodied by Messrs Offord & Sons for the Delhi Durbar of 1911.*

Below: *A 1912 30hp cabriolet, superbly executed by Arthur Mulliner of Northampton.*

Left:
An impressive 57hp Daimler, although a more flowing line applied to the wings would greatly add to the appeal of the car.

Below: *A Daimler 15hp tourer dating from 1912.*

unit. In ordinary light running, the two engines drove the road wheels, and at the same time, the dynamotors acted as dynamos charging the accumulator situated underneath the driver's seat. When an additional load had to be taken, the accumulators automatically came into play, driving the dynamotors and changing them into motors, thus increasing the vehicles available power; either power unit could be changed without the need to remove a body panel. Other innivative features included four wheel drive (which was still virtually unknown on private cars of the time) and the mounting of the engine amidships.

During the same year, the old London General Omnibus Company were manufacturing bus chassis at their Walthamstow factory. Something like 3,000 chassis' were turned out, and the Daimler Motor Company supplied the firm with many sleeve valve engines to power them. By 1912, the association which

Right:
A cross section of the Daimler models for the 1912 trading season.

had been growing between the two companies resulted in the formation of a concern known as the Associated Equipment Company (AEC), allowing for the manufacture and purchase of components which could then be assembled at the London works.

The Merger between BSA and Daimler

To return to the amalgamation with BSA, the first newspaper to state that something important was going to happen was the *Financial Times.* On 2nd September 1910, they reported, even before the Principal Agreement had been signed, that *"negotiations were proceeding for an amalgamation of the BSA and the Daimler Motor Co. (1904) Ltd of Coventry"*. The terms were awaited with keen interest in the financial world, as it was realised that this would be by far the most important merger that had yet taken place in the British motor industry.

Left:
A typical shooting brake for the period. This example was built for the Hon. John Ward.

Above: *A BSA tourer in Singapore, pictured shorly before the outbreak of the Great War.*

Below: *Built for Lord Lilford of Northamptonshire, this 1912 30hp limousine carries Daimler's own "Earlham" coachwork.*

Right:
A 38hp Special Daimler as introduced in 1912 with a Compton six seater body.

Below: *A rare picture showing the Daimler stand at a typical foreign pre-WW1 Motor Show.*

When published, it was noted that Daimler would not lose its identity, but the existing Company would be wound up, and a new one established with BSA holding all the shares (registered as company number 112569, the Daimler Company Limited was incorporated on 7th November 1910 with a share capital of £250,000). BSA was in a very sound and prosperous position, with consistently large profits and an authorised capital of no less than £1,000,000 - they paid Daimler's ordinary shareholders 18% on their investment, and in effect paid just over £600,000 for the company.

Above: *A beautiful sporting Daimler from circa 1913. A similar 38hp car was supplied to the Maharaja of Barwani.*

Below: *A Royal 57hp limousine at Houghton Towers with King George V and Queen Mary, pictured in 1913.*

Above: *A 20hp four cylinder Daimler car of 1913, seen here within half a mile of the works.*

After the first AGM following the amalgamation, held at the Grand Hotel in Birmingham on 4th October 1911, the new Daimler Board was announced as follows: Sir Hallewell Rogers JP, Lincoln Chandler, Frank Dudley Docker (the architect behind the merger) and Albert Eadie were elected as ex-BSA Directors, whilst Sir Edward Manville (Chairman), Arthur Herbert Wood, and Percy Martin (Managing Director), were named as the ex-Daimler Directors. Ernest Instone continued to be the Manager of the reconstituted Daimler Motor Company, and it is interesting to note that after their first year, the profit of both companies amounted to well over £200,000.

Daimler in France

The troubled history of the Daimler Company in France was further aggravated by the loss of Loén Girardot at the end of 1911. By 1912, he was building the "Girardot" car. Based on the Charron chassis (Charron had bought out CGV, Girardot's first car manufacturing concern) it used a Panhard sleeve-valve engine, the patents for which belonged, rather ironically, to Knight-Daimler.

Nearer to home, J.D. Siddeley had negotiated the

Above: *A rubber tyred Daimler lorry from 1913 in the livery of Davenports brewery.*

Below: *A 1913 30hp Camborne limousine by Daimler's own coachbuilding shop. The four cylinder model ranged in price from £805 to £870; the six cylinder engined car cost an extra £65.*

supply of sleeve-valve engines to Deasy (soon to be Siddeley-Deasy). These were used on the larger vehicles of the marque, and signified a continuation of Siddeley's link with the Daimler Company. Indeed, at the time of the First World War, there was talk about a possible merger between Siddeley-Deasy and Daimler, but these came to nothing. Before this, there was another link, however.

BSA Motor Cars

The BSA Company decided that the time was right for them to re-enter car manufacturing. It is widely believed that BSA had never built a car prior to the BSA-Daimler merger but, as mentioned earlier, they had already made some tentative steps into this field. In fact, by October 1908, BSA were producing three different models; a 14/18hp, an 18/23hp, and a 25/33hp - at least two of these were exhibited at the Motor Show of that year, promoting them for the 1909 season.

With this in mind, it would be more correct to say that BSA never really made much of an impression on the motor industry until the amalgamation with Daimler. Dudley Docker blamed this lack of success squarely at Col. Ernest Baguley, the man in charge of BSA's car building operations, going so far as to say he was *"incapable of organisation."* Baguley left BSA after the take-over to build his own cars (known as the Baguley), in Burton-on-Trent.

Once the merger had taken place, no time was lost in producing a new BSA car with several innovative features. The car, available only through the Company's appointed agencies, was first exhibited at the 1911 Motor Show. This 1912 model BSA had a Knight sleeve-valve engine rated at 11.9 hp, and a pressed steel body produced at Sparkbrook. In fact, the car looked so similar to the Stoneleigh, which was marketed by Siddeley-Deasy during the same period, that it has been said they were one and the same, just with different badges. When the Great War broke out, the BSA model was phased out, and perhaps thankfully not revived.

Below: *Spring wheels trials being carried out on a 20hp Daimler. The Company had been experimenting with wheel types for a number of years before the war.*

Left:
A 38hp Special Daimler chassis carrying a Mann Egerton body. This car was built for the HM the Queen of Spain in September 1914.

Right:
A Daimler 38hp tourer by Mulliner, built between 1913 and 1914.

Left:
A Daimler 45hp Special chassis with an unusual body by Arthur Mulliner. It was built to order for Mr Cassamally Peerbhoy, a leader of the Muslim faith and a notable car enthusiast, from India.

Above: *A 1914 Daimler coach, seen here just managing to squeeze through Coventry's Cook Sreet Gate.*

The 1912 Daimler Range

For 1912, the Daimler Motor Company decided to produce seven different models using four and six cylinder Knight engines, which by now had been modified and were nearing perfection. The six cylinder 38hp was said to be so smooth and free from vibration, that it felt as if *"one is being towed by an invisible cable"*. The 25hp model also gained some wonderful write-ups, the *Autocar* noting that *"the cars running is superlative."*

In all but the 15hp and 23hp models, the external connection between the cylinder and the water jacket was done away with in favour of the, now common, internal connection. There had been a rumour that the Company were going to abandon the sleeve-valve altogether for the 1912 range, but this was totally unfounded.

By this time, the sleeve valve (or slide valve as some put it) had become well established. *The Autocar* noted that, in their experience of the 15hp models, it had a number of advantages -
"Its from troubles arising from dirty valves, cylinder heads or pistons. Its ability to stand up to its load and its quietness at low and medium speeds, this quietness being maintained without any necessity for adjustment."

They also pointed out some disadvantages. *"more vibration and noise at high engine speeds, and the inability to restart on the switch,"* but added *"we believe that one, if not both, has been overcome."*

During 1912, Dudley Docker resigned from his seat on the BSA Board, to be represented by his good friend Lincoln Chandler. In 1913, however, a noteworthy addition was Neville Chamberlain (Britain's Prime Minister from 1937 to 1940) whose uncle, Herbert Chamberlain, had been the Chairman of BSA for several years. Despite his political powers though, many of Neville Chamberlain's colleagues felt he was *"an indifferent businessman"*.

He probably wouldn't have been offered the position had it not been for his father and his life long friend, Arthur Wood, who was a Daimler Director from 1904, and a BSA Director from 1907. Arthur Wood stayed on both Board's until his death in 1934, although he too was noted for his lack of business judgement, being infinitely better on the the river bank than in the Boardroom. Wood had been given his seat simply because he was one of the largest Daimler shareholders. Colonel Henry Holden (who later became Brigadier-General Sir H.C.L. Holden KCB, FRS, and was the Chairman of the Royal Automobile Club for

Above: *The Daimler Chassis Erection Shop shortly before war was declared.*

Above: *The King of Tunis pictured with his 15hp landaulette in 1915, by which time the car had already been in service for a number of years.*

Below: *A magnificent TL.20 landaulette dating from 1914, chassis number 12659.*

Above: *A Maythorn design proposal for a three quarter cabriolet on the Daimler 45hp Special chassis.*

1904/05), also became a Director, as did Thomas Ferdinand Walker (a BSA Director from 1887 to 1919).

The Company gradually began to move many of its operations to the Coventry Radford site it had bought about five years earlier. Christened the "Radford Works", over the next few years it would gradually become more important than the old Motor Mills.

The Last Pre-War Machines

The highly-successful 15hp was discontinued for 1913, as was the 23hp model. Only the original four cylinder 38hp remained unaltered from the previous years range, and it was joined by 20hp, 26hp and 30hp cars. In addition, there was the new "Special Series" 38hp chassis.

The 20hp (90mm x 130mm) was the smallest Daimler for 1913, and was basically the same as the 1912 version except for better engine lubrication. The 26hp was just a slightly enlarged version of the 25hp from the previous year (101.5mm x 140mm as opposed to 101 x 130mm), and the 30hp car was basically a six cylinder edition of the four cylinder 20hp described above. Standardised models were becoming the norm at this time, so when Daimler announced their six cylinder "Special Series" 38hp model, it caused considerable interest. Using the same bore and stroke as the four cylinder 26hp chassis, this latest car was a genuine de luxe vehicle.

For the following year, the range was cut to just four chassis types; a redesigned four cylinder 20hp model, a four and six cylinder version of the 30hp, and a large 7.4litre 45hp Special, which replaced the 38hp Special Series in the new line-up.

During this time, there were a bewildering amount of body styles available. The Company itself listed around fifty distinctive designs and these, taken in conjunction with the number of different chassis types and the many individual coachbuilders who could offer suitable coachwork, gave the Daimler customer an enormous choice.

One of Daimler's most interesting projects from the pre-war period, was a petrol-electric railcoach. In October 1913, it was reported that the coach had achieved a speed of some 70mph on the fastest part of the line stretching from Coventry to Kenilworth. This was probably the highest speed achieved by a train at that time, as the steam trains of the day were rarely able to exceed 40mph.

Agricultural tractors were turned out in large numbers immediately before the First World War, and many were purchased by the Government when the question of stimulating farming became so pressing. During the war itself, these tractors were often used for hauling large field guns into position.

Below: *In the foundry at the Daimler works. Over the years that followed this particular part of the factory would prove to be extremely useful.*

Chapter Five

DAIMLER GOES TO WAR

From its beginnings until the outbreak of the First World War, the development of the internal combustion engine had been steady, with new achievements being developed to meet market demands. The war provided a new stimulus and accelerated power unit development enormously. Vast numbers of motor vehicles of all types were needed, and in addition, the aeroplane, which came into its own during the war, was still only in its infancy. Prior to the outbreak of war, Daimler had been producing around 25 commercial chassis per week, but due to the demands of the WD this figure was soon doubled.

Before the First World War, the most popular heavy Daimlers were two to five tonners with 4.0litre to 5.7litre petrol engines, cone clutches and flitch plate frames. The experience gained by the Company in the years leading up to the conflict, the reliability of its products, and the fact that they could continue working with the minimum of servicing, resulted in the authorities ordering large numbers of Daimler commercials for use in all theatres of war.

During the first few months of the war, the Company was inundated with fresh demands from the War Department. Fifty wagonettes were delivered by October 1914, and in the meantime, work had began on 20hp staff cars with bodies that doubled up as sleeping berths for the Officers; a total of 150 were produced.

Facing page: *Part of the shell making operation at Radford. The arches in the wall can still be seen today.*

The last Daimler Car delivered before the War.

Left:
This was the last car to be delivered before warwork took over production. A 45hp Special Daimler "Sandford" phaeton finished in grey and black.

Below centre:
A Daimler ambulance parked in Paris during the early days of the war.

The first type of Box Body fitted to Daimler 20 H.P. Chassis

Left:
A 20hp Daimler chassis fitted with a typical wartime box body.

Above: *A Daimler staff car. Most of them were converted limousines or landaulettes which seated two people in the front and two more in the rear compartment which had folding tables.*

War was declared on 4th August 1914, just a few days after the annual works holiday had begun, so very few members of the workforce were in Coventry. Over 200 employees were reservists or territorial, so they reported directly to their regiments, while gradually the Daimler staff was brought back to the works.

In the early days of the conflict, almost any vehicle possessing four wheels and an engine was ruthlessly requisitioned. Almost immediately, the War Office stripped the factory of over 100 cars and lorries; even the graceful limousines and landaulettes were requisitioned, with bright metal fittings being painted khaki. During the first few weeks, sixty Daimler commercial vehicles of two, three, four and five ton capacity which had been ordered and were awaiting delivery, were taken over and despatched to Avonmouth for shipment abroad. Daimler buses, delivery vans and touring cars were used for carrying goods. As if this wasn't enough, instructions were sent to every Daimler depot in the country to send every vehicle it had to a marshalling point in Hyde Park. Soon, around 200 vehicles were gathered, and duly taken over by the military authorities.

The War Office urgently required a large number of 20hp chassis fitted with box bodies. The design, although basic, was a totally new one, so had to be submitted for approval first, but once this was done, some sixty cars were supplied in just eleven days. Daimler ambulances were also in great demand, and King George and Queen Mary donated a pair on the advise that the Daimler machines were the most suitable in the field.

An interesting development was the use of Daimler three ton trucks as travelling workshops. The solid frame sided lorry body with its hinged doors and tailboard, was fitted with leg supports, so that when opened up, a large amount of floor space was made available. Inside, there was a 6 inch lathe, a wet tool grinder, a small bench drill, a bench with two vices, a portable forge and a set of tools. These mobile workshops were attached to convoys of lorries, and allowed repairs to be executed on the spot.

Another new vehicle was a limousine with the driver seated outside for security - 600 were built,

Above: *This unusual vehicle is a Marconi Wireless van.*

Below: *One of the hundreds of mobile workshops built by the Company for use in the field.*

Right: *The first Daimler built Gnome Rhône aero engine on test in September 1914.*

Hangars on the Daimler Aerodrome.

establishing new production records for the Company in the process.

One of the most remarkable achievements concerned the fulfilment of a sudden order placed for 50 two ton machines, placed during the first week of March 1915 on the condition that all of the vehicles should be delivered to Avonmouth by the following Saturday. After five days and nights of supreme effort, the order was completed. It was learnt subsequently that the urgency was caused by the fact that they were needed for the Gallipoli landings. Once again, the Daimler Motor Company was highly commended.

Daimler in the Air

Britain was considerably handicapped during the first couple of years of the war due to a lack of suitable aero-engines - those that did exist were usually French, and very rare in this country.

One of these engines was the 80hp Gnome rotatry engine, with its seven revolving air-cooled cylinders on a stationary crankshaft. The British motor industry had no experience of such an engine; the Daimler engineers having never even seen one, nonetheless, the authorities approached the Company with a view to manufacturing it.

There was not a jig or tool, gauge or stamping for the Gnome, but by the 7th August 1914, arrangements for the Daimler Company to produce it were concluded. Precisely how this was going to be done, no one quite knew.

An engine was delivered from Hendon to the Daimler works, and on its arrival it was completely dismantled in the Drawing Office. Work continued on the engine day and night, with the result that within a week, every single component had been measured and drawn. An illustrated specification was compiled,

Above upper: *A steam tractor pulling a B.E.2c aircraft crate.*

Above lower: *Daimler's first aeroplane prior to take off in June 1915.*

printed and issued to each of the various departments involved.

Special materials were needed, and meeting this requirement was helped by the analysts of the National Physical Laboratory. Problem after problem cropped up, due mainly to the lack of information and experience in building aero-engines. As time went on, it seemed unlikely that the Company would be able to keep its promise of having the first Gnome up and running within eight weeks. Gradually, Daimler's efforts began to come together; a special test stand was acquired from Gnome-Rhône themselves, and a propeller was requested on the final day. Unfortunately, this went missing, and did not arrive until after dark which, considering that the test was to be held outdoors, was a somewhat awkward mishap.

On the last day of September, however, the engine, complete with propeller and illuminated only

Above: *Four types of aircraft were built at Daimler's Radford works. First the B.E.2c, then the B.E.12, the R.E.8 and finally the D.H.10 bomber.*

Below: *The largest aircraft ever built at Daimler, the D.H.10 bomber. This picture shows the first of the type at the Radford works.*

by a car headlamp, stood ready to be started for its initial test. After a few last minute difficulties, it burst into life amid great excitement on the part of all those involved. Incredibly, the Company had fulfilled its promise with exactly one hour to spare.

Another aero-engine built by Daimler, and based on the Renault 70hp, was the RAF.1. The contract for this engine had been put out to tender by the Government, and when the Company learnt that they had secured the order on the 30th August, work began in the Drawing Office on the following day. This was later developed into the 90hp RAF.1A, and the RAF.4A (140hp) was another Daimler production.

Contracts for the Hispano-Suiza and Sunbeam aero-engines came to nothing, but towards the end of the war, no fewer than 1,245 W.O. Bentley-designed BR.2 nine cylinder aero-engines were also turned out, the original order (which would have taken the total to 2,000) being cut short by the Armistice.

One of the biggest difficulties encountered by the Coventry firm was in the testing of its aero-engines. Being air-cooled, the tests were best carried out using the facilities in the Royal Aircraft Factory, but it soon became evident that this was not an ideal arrangement. Eventually, a large propeller fan enclosed in a cage was used, and after a little experimentation, was found perfectly adequate to allow Daimler to do their own testing at the works.

Shortly after, it was suggested that the Company should start building complete aircraft - a proposal which led to the creation of a whole new department, as well as the building of a new shop for their construction. The idea was not totally new to the Company, however, as in January 1911 it was noted: *"The Chairman submitted details of Mr Holt Thomas' proposal for the building and sale of aeroplanes, but it was decided to defer entering into this class of business for the present."*

While all the necessary material was being ordered, two foremen and an engineer were selected to go to the Royal Aircraft Factory for training. This department soon expanded out of all recognition, and by the time the shop was ready for occupation, it proved to be much too small, prompting the Company to build yet another bigger one. Further expansion also took place in the aero-engine section of the factory, all of which resulted in the workforce increasing from just under 4,000 to well over 6,000 people, including a large number of women as the war drew on.

With aircraft production now well in hand, the War Office looked at the possibilities of building another aerodrome in the Coventry area. Officers were sent to look at the land, and it was decided to utilise the Company's Radford Estate. This site had been purchased by the Daimler Motor Company some years ago, and was placed at the disposal of the Government for the period of the war. However, the War Office decided to enlarge the whole scheme, and the aerodrome was subsequently placed under the control of the newly formed Royal Air Force. Daimler received the sum of £1 per annum rent for the land, paid by the Secretary of State. The complete aircraft built at Daimler (including their Serial Numbers and type of engine) were as follows:

B.E.2c	2030-2071 (RAF.1A engines)
B.E.2c	2072 (Renault engine)
B.E.2c	2073-2121 (RAF.1A)
B.E.2c	2122 (Renault)
B.E.2c	2123-2129 (RAF.1A)
B.E.2c	2570-2598 (RAF.1A)
B.E.2c	2599 (Hispano-Suiza)
B.E.2c	2600-2669 (RAF.1A)
B.E.12	6478-6510 (RAF.4A)
B.E.12a	6511 (RAF.4A)
B.E.12	6512-6677 (RAF.4A)
R.E.8	A3531-A3541 (RAF.4A)
R.E.9	A3542 (Sunbeam)
R.E.8	A3543-A3560 (RAF.4A)
R.E.9	A3561 (Sunbeam)
R.E.8	A3562-A3680 (RAF.4A)
B.E.12a	A4006-A4055 (RAF.4A)
R.E.8	A4161-A4260 (RAF.4A)
B.E.12	A6301-A6302 (RAF.4A)
B.E.12a	A6303 (RAF.4A)
B.E.12	A6304-A6336 (RAF.4A)
B.E.12a	A6337 (RAF.4A)
B.E.12	A6338 (RAF.4A)
B.E.12a	A6339 (RAF.4A)
B.E.12	A6340-A6350 (RAF.4A)
R.E.8	B3401-B3450 (RAF.4A)
R.E.8	B5001-B5150 (RAF.4A)
R.E.8	C2231-C3080 (RAF.4A)
B.E.12a/b	C3081-C3280 (RAF.4A)
D.H.10	E9057-E9206 (See below)
R.E.8	F3548-F3747 (RAF.4A)

Totals for each aircraft type are as follows:

B.E.2c	200 off for the Army.
B.E.12/a/b	500 off for the Army/RFC.
R.E.8	1,500 off for the RFC/RAF.

Above: *Daimler 105hp tank engines awaiting despatch.*

Below: *The original British tank was powered by a 105hp engine, the machine's outline can just be seen. On later models, the transmission was also built by Daimler.*

The D.H.10 bomber, powered by either two 400hp Liberty engines or two 360hp Rolls-Royce Eagles, should have had a run of 150 (ordered by the RAF), but it is thought that a figure nearer to twelve is correct due to the end of the war. Although the last batch of R.E.8s were built after this, they were considered more useful and put into storage, although the total here may well be 1,390 instead of 1,500.

In view of the Company's Stirling work, Percy Martin, Daimler's Managing Director, was made Controller of Petrol Engines under the Ministry of Munitions and also given a seat on the Air Board. Later on, the Company's Chief Engineer, A.E. Berriman, was appointed Deputy Controller of the Technical Department by the Ministry - a position he held until the end of the war. Incidentally, the Assistant Chief Engineer at this time was G.T. Smith-Clarke, who became the Chief Engineer at Alvis in 1922.

Daimler and the Tank

Tanks were developed during the Great War from the Daimler-Foster tractors which were designed to pull siege guns over difficult terrain. In actual fact, the original tanks came about as a result of a portable bridge invented by William Foster & Co. to get the tractors over ditches. Daimler engines were used in these, so when the idea of the tank finally came to fruition, Daimler engines and transmissions were a natural choice for the job. The engines were six cylinder Knight sleeve valve units of 105hp, and were designed to produce this output at a very low 1,000rpm. After a certain amount of development work by Harry Ricardo, the Daimler engines proved themselves to be thoroughly efficient in action, unlike the armour on the earlier machines.

Tanks were first used during the Battle of the Somme, and were subsequently employed in the Cambrai attack where their value became apparent. These early tanks were fitted with a large rear wheel for steering purposes, but this proved to be totally unsatisfactory as the wheels were almost wholly exposed and were often shot to pieces. Towards the end of the war, the Company designed a new system of petrol-electric transmission which allowed the rear wheel to be abandoned. By using two engines, each transmitting power separately through an electric motor to the tracks, the tank could be steered and made to revolve almost on its own axis. This method of transmission was later used on several experimental Daimler projects.

By the end of the war, the total output of Daimler commercial vehicles far exceeded 4,000 units. A vast number of shells were also produced by the Daimler works, and on the 22nd June 1915, King George V paid a visit to the factory. Various departments were visited, before he went on the BSA factory in Birmingham. Finally, on visiting the front, the Prince of Wales is said to have remarked, *"It seems to me that the Daimler people are running this war!"*

Daimler's war effort was indeed magnificent, and was only bettered by the work carried out by the Company during the Second World War, but there was no denying the conflict had been good for them financially.

At the end of 1913, share capital was increased to £550,000, but in May 1918, they were able to increase this figure by a further £1,000,000. For the first time since the merger, BSA would at last be able to finance Daimler projects properly.

COVENTRY ELECTION
SUPPLEMENT TO THE
LABOUR LEADER
No. 50. Vol. 15. DECEMBER 12, 1918 ONE PENNY

MANVILLE MINOR: "Say, dad, what did **you** do in the Great War?"
B.S.A. MANVILLE: "Oh, I helped the B.S.A. to make £435,207 profit during 1917."
MANVILLE MINOR: "Yes, dad, but I said **what** did you do, not **who**?"

Above:
The Labour Party released this cartoon of Sir Edward Manville in December 1918 at the time of the local elections. Manville was standing for the Tory party.

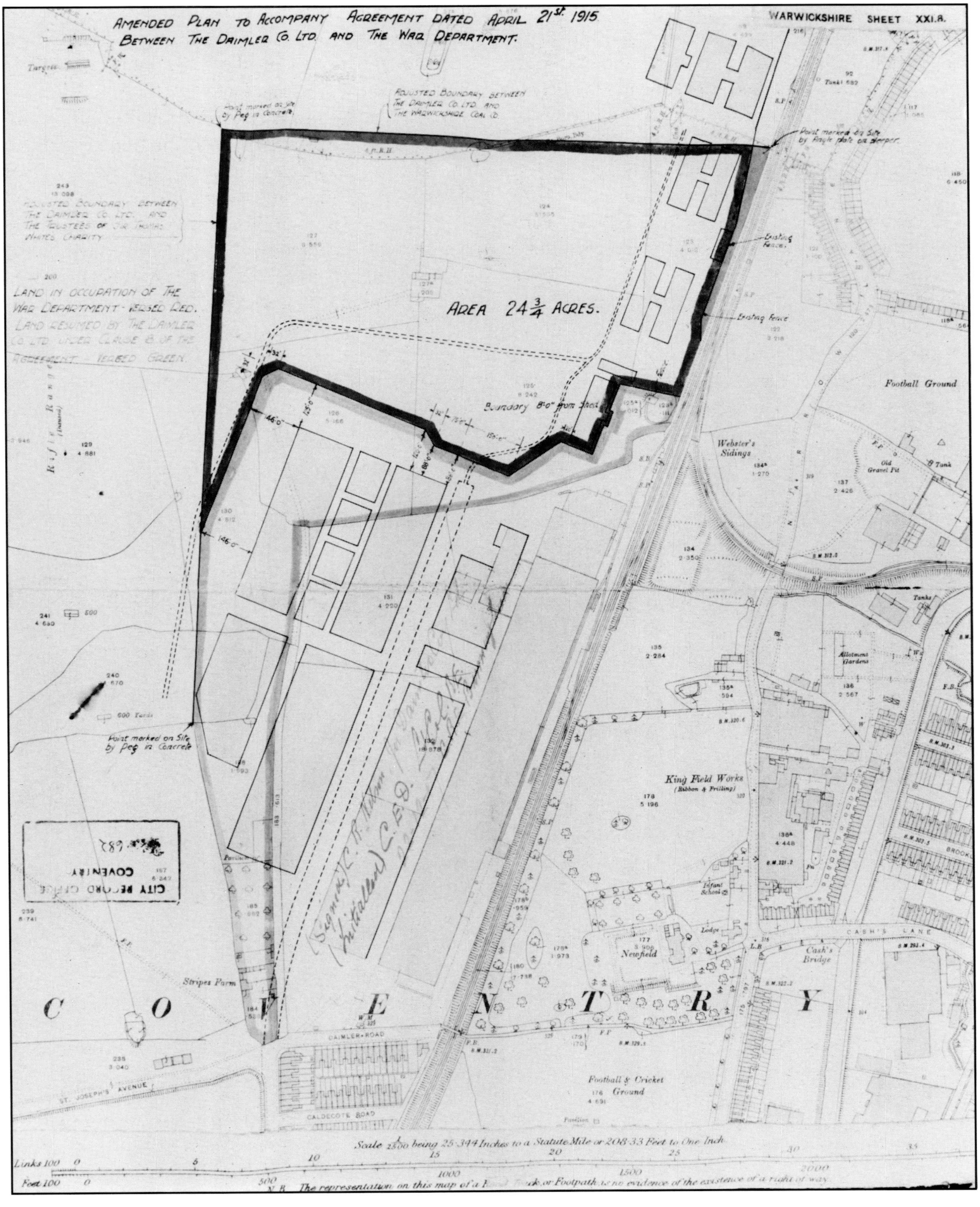

Above: *Map showing the site of the Radford Aerodrome.*

Chapter Six

THE VINTAGE YEARS

When the war finally came to an end in November 1918, the British motor industry found itself in a difficult position. Four years before, mass production methods were being considered by several of the country's manufacturers after Henry Ford had shown the way forward, but the hostilities that had flared up in Europe had interrupted these plans.

At the end of the war, virtually every manufacturer found itself left with a huge amount of worn-out machinery, often housed in hastily erected temporary workshops. A great number of employees were thrown out of work, and there were no fixed production schedules for the future.

On the other hand, during the war years, there had been a great change in public opinion towards the motor car. Hundreds of thousands of people had been taught how to drive, and to the man in the street, the automobile was now a familiar object. Vast numbers of recently demobilised men, many of whom had just been awarded their army gratuities, were now free to start enjoying a new life, thus bringing about a minor trade boom. Demand was greater than supply, and veteran cars which had been run into the ground on war service managed to fetch higher prices than their original cost.

However, this situation soon came to an end with the great trade slump of 1920.

Facing Page: *The typical splendour of the rear compartment of a 1920s Daimler limousine.*

Above: *With the war over the Company could get back to building high quality motor cars.*

Below: *A six cylinder Daimler on power test on a Froude dynamometer. Note that power is being measured at the wheels. The 45hp car should have given 65bhp and the 30hp at least 45bhp.*

Left:
A special limousine dating from around 1919 mounted on a 45hp chassis.

Daimler had not really considered mass production, claiming that the quality needed for a car bearing the famous name would be compromised. This decision, if nothing else, helped the Company to recommence production a little sooner. The announcement by the Society of Motor Manufacturers & Traders in the early part of 1919 that a Motor Show would be held at Olympia in November, gave every manufacturer something to aim for. In the meantime, the Company carried on producing the six-cylinder cars in the 1915 model range, which were basically the same as those in the 1914 range.

The 1920 Model Year

It was decided that the Daimler programme for 1920 should be the same as that for 1919, and consist of two 30hp models; the "Light" and the "Standard" Thirty, and a special 45hp production. Both the Light and Standard Thirty's were fitted with a six cylinder sleeve valve engine of some 4,962cc. There was at least one interesting and innovative feature of this power unit - the lubrication was semi-automatic, operated by scoops and troughs beneath each connecting rod. These were interconnected with the accelerator so that the scoops dipped into the troughs more frequently as engine speed rose. There was also a vibration damper on the front of the crankshaft which proved to be very effective.

Little else was new in the cars design, except that three-quarter elliptic springs were used on the rear suspension - a feature also adopted on the 45hp model. Drive was through a leather-faced cone-clutch to a four-speed gearbox via two leather couplings. The Light Thirty in effect replaced the 20hp model, being slightly lighter and shorter than the Standard model, and returned an average of 18 mpg because of this. The Light Thirty was sold for £1,060 in chassis form. Alternatively, the car could be bought with a standard Daimler-built body for just £1,610. When these first post-war models were announced, a special point was made of the fact that they would be fitted with an electric starting and lighting set, as well as a speedometer.

This new range of vehicles were given the designations Type TG for the Special 45hp with an upright steering column, and TH for an equivalent model with an inclined column. The standard 30hp was given the references TL and TM respectively, while the Light Thirty was known as Type TO. The Type TP being a Light Thirty with a raked steering column, but this was available as a special order only.

The first post-war Motor Show opened on 7th November 1919. The Daimler Board had decided that henceforth, the Company's products would be sold through the motor trade, and not direct from the works as had been the previous arrangement. It was noted at the Show that a Daimler 45hp Special limousine cost around £2,100, while a Rolls-Royce Silver Ghost chassis alone cost £1,575. Several coachbuilders displayed their art on the Daimler, including Salmons, William Cole, Hamshaw, Hoopers

Above:
The pre-war petrol electric rail coach pictured in 1919.

Right:
A Hooper tourer on the Daimler Light Thirty chassis circa 1921. This car was delivered to the Duchy of Cornwall.

and Windovers. In order to help the coachbuilding trade, Daimler had patented a steel bodyframe for their larger models, which in due course could be matched up to the chassis. By doing this, the coachbuilder could start on the body, knowing it would be a good fit, without the need to wait for a completed chassis, thus saving a considerable amount of time.

The Company had by now one of the finest metallurgical departments in the world, and this was instrumental in maintaining high standards of quality. With the same aim, the Daimler toolroom produced equipment of such precision and accuracy, that master gauges emanating from the Coventry works were found throughout the industry. During 1920, the Company filed a patent (number 139116) for an easy-starting system developed by Algernon Berriman. It was never taken further, however.

Daimler Hire

From 1920 onwards, every effort was made to increase the efficiency of the repair depots and Daimler service stations throughout the country, and to add to their number, so that a chain of these could be established for the benefit of the Daimler owner. It was also realised that the demand for the hire of top class cars had so far only been met in a confused and unsatisfactory manner - there were certain sections of the public who often required the use of a large chauffeur-driven car from time to time. Although Daimler had been doing this in London in a small way for perhaps 10 to 15 years, to try and meet this requirement fully, a new company called Daimler Hire Ltd was founded with around 250 of the latest Daimler cars at their disposal. Based at the Prince's Skating Rink at Knightsbridge, the business proved to be a great success for many years.

In 1923, Daimler Hire was awarded the Royal Warrant by King George V, and their vehicles could be seen at numerous official and State functions; the Company is now in the hands of Hertz. An interesting offshoot from this distinguished hire firm was the business known as the Daimler Airways, formed in April 1922.

Daimler Hire's parent company, BSA, had bought the Aircraft Manufacturing Co. in 1920, and with it came the subsidiary Air Transport and Travel. At this time, however, there was no Government incentive, and many of the early airlines found life difficult. Indeed, BSA found that Air Transport and

Left:
Although the basic charabanc design had changed little when compared with the pre-war models it did feature such luxuries as pneumatic tyres. The bodies also had a far cleaner, smoother line.

Right:
A pair of cars built for an Indian Maharaja around 1921. The Maythorn bodies were built on the 45hp chassis. During this period many cars were shipped to India. It has been said that 25% of Rolls-Royce production was also exported to India at that time.

Right:
A Light Thirty landaulette, by Knibbs of Manchester.

Below:
Another Light Thirty, this time with a Hooper saloon body, circa 1921.

Travel was making a heavy loss, and decided to close the operation down. This experience gave the firm an insight into the world of air transport, and after a subsidiary scheme was introduced in 1921, Daimler Airways was founded. The Company flew to several European cities using de Havilland D.H.34s, but lasted only until April 1924, then, like many of the independent airlines, it was absorbed into the Imperial Airways concern. Shipping was also an important part of the business, as suggested by the Daimler Hire motto; *"Hire by Land, Sea and Air."*

The Company owned several ex-Royal Navy MTBs, which were converted into luxury cruisers powered by Daimler 105hp tank engines. The firm also ran a Continental coach service, each with its own toilet and coffee bar, and made available the facility to hire the many different types of Daimler commercial vehicles.

A new Policy

Despite the ever-increasing threat of industrial action, in January 1921 it was announced that due to the increasing popularity of Daimler cars, the Company was able to reduce the prices of certain chassis. The Light Thirty was now £900 instead of £1,125, while the standard Thirty was adjusted to £1,000 from £1,150. The 45hp Special chassis, however, remained at £1,450 until much later in the year, when it dropped to £1,275. Daimler still listed a large number of standard body styles, and these included the Castleton and Clifton

Above: *A delightful poster produced by Daimler Hire. Note the Daimler Company's Paris Showrooms being depicted as the French offices for their operations on the Continent.*

Right:
Headquarters of the Daimler Hire organisation in Knightsbridge, London.

Below:
This Light Thirty tourer was apparently delivered to Emperor Hirohito of Japan. A number of cars went to Japan in the vintage period, some being bodied locally by the Japan Automobile Co. of Tokyo.

tourers; Chesterfield, Mirfield, Wingfield and Southfield landaulettes, and the Saltley and Berkeley limousines, some of which were available in de luxe form. The most expensive of these was the 45hp Special Berkeley limousine, which sold for £2,000 after the price review.

Instead of waiting as usual for the Motor Show in October, the Company decided to announce its latest new model - the TT4-20, in July 1921. This was to be the last four-cylinder car Daimler ever produced, although some of the Daimler-built Lanchesters constructed later on in the firm's history did sport this engine configuration, as did some of the BSAs. To return to the TT4-20; it was a sleeve valve 3.3 litre 20hp model, built along similar lines to its larger six cylinder brothers. It had an 11 foot wheelbase, with adjustable rear wheel braking. The chassis cost £750, and was highly-respected according to contemporary reports.

During the latter part of 1921, Ernest Instone resigned from his seat on the Daimler Board. He had been with the Company for nearly all of its life, but happily, his severance was not a complete one, as in partnership with the firm's London Manager, Undecimus Stratton, he formed a company known as

Stratton-Instone Ltd. This new business assumed the command of the London Daimler premises at 27 Pall Mall, and handled sales of all Daimler, BSA and, at a later stage, Lanchester cars, in the London area. For many years, Stratton-Instone (or Stratstone Ltd, as the firm became known after Ernest Instone died) had the honour of attending to all the motoring requirements of the Royal Family. Stratstones are still trading in Daimler cars today, and under the direction of James Smillie, have recently expanded their operations in the capital.

The 1922 Range

The Daimler range for 1922 was kept much the same until July, when a 2lhp six cylinder car was announced. This was similar in most respects to the model it replaced - the short lived TT4-20, except for the engine. With a bore and stroke of 75mm x 114mm, the capacity was just over three-litres in a monobloc engine with detachable cylinder heads. The 30hp model was redesignated the TS6-30, while the 45hp had become the TJ6-45 (Special).

A new BSA light car was also introduced in 1922, but it was soon replaced by a series of sleeve valve models which were basically just smaller versions of the current Daimlers. These were built at the Coventry factory, allowing BSA's Sparkbrook works to be turned over to the manufacture of machine tools. In 1926, BSA stopped building cars, and for the next three years there were no BSA cars on the market whatsoever.

Daimler and Marconi

The first experiments to equip motor cars with wireless - then the latest product of science - were made on Daimler cars. During the summer of 1922, the Company had been in touch with the Marconiphone Company with a view to some experimental work being carried out in time for the forthcoming Motor Show at Olympia, and wireless apparatus was duly fitted to a few cars. It was installed underneath the front seats with a folding frame aerial mounted on the roof. The equipment consisted of several tuning circuits with a seven volt aperiodic amplifier, which had been developed during the war. The output was taken to telephones, and in order to obtain satisfactory results, the entire ignition system of the car had to be carefully screened by using braided cable and metal clad plugs, again developed during the war for use in aircraft.

Taking into consideration the speed with which the work had been completed, and the primitive state of both broadcasting and reception at the time, the results obtained were surprisingly good. The success was due largely to the quietness of the cars, but there was obviously a need for further development. The aerial was both fragile and unsightly, and badly adapted for speed. The screening presented a drawback as well,

Below: *A Daimler Hire 45hp sleeve valve limousine pictured in Ireland during July 1925. The chauffeur is Mr Degonda who was with the Company for a number of years.*

Above: *The early days of Marconi radio. The car is a TT4-20 of around 1922.*

Below: *This Daimler, dating from 1922, has been fitted with a superb Barker Salamanca style body.*

Above: *This 1923 16hp Doctors coupé with dickey seat was bodied by Shaw.*

as it reduced the performance of the engine.

After the event at Olympia, the cars were taken to the Scottish Motor Show held in January 1923. On arrival, it was found that the transmission from London was too weak to be picked up in Scotland, so for the duration of the Show, the Daimler Co. obtained a licence to set up a temporary transmitting station. In these early days, however, the public was not particularly attracted by car radio. Something like a dozen of the cars used by the Daimler Hire fleet were fitted with radios, but there seemed to be very little interest in those particular vehicles. The drivers were also not keen on the idea; they seemed to think that the wireless added to their responsibilities.

The experiment continued for a year, but as the demand gradually dwindled further, the cars were dismantled. The whole enterprise was really ahead of its time, the wireless in the home was still something new, and so wireless for the motor car was perhaps just too innovative. During 1924, another effort was made to popularise the car radio through the means of a portable set. Once again, there was little success, and the project was abandoned.

In March 1923, there were a number of new arrivals on the Daimler Board. Richard Rotherham from BSA's French company was one, and Willoughby Bayley from BSA Cycles was another, but Walter Shephard was perhaps the most interesting, as he was later the Managing Director of Tillings-Stevens, the Group that bought Daimler Hire from BSA in 1930.

More New Cars

There was an interesting 12hp Daimler car produced for 1923, which had a six cylinder sleeve valve engine of just 1,542cc. It had been rumoured that the model was on its way in April 1922, but it took until November for it to make its debut. At the same time, a 16hp machine of just under 2.2litres was announced, showing signs that Daimler was aiming to cater for a far wider market. In fact, no fewer than 56 different chassis and body combinations were available in 1923 (including BSAs). In addition, there was a huge extra choice of bodies available from individual coachbuilders. The cost of a 12hp chassis was put at £550, with the 16hp version priced at £625.

Author, George D. Abraham, took one of the new cars into the Lake District, and noted, *"The Daimler car, a good sample of the new 16hp model, with the six-cylinder Knight engine, seemed to feel the joy of the morning. Its noiseless flight, so typical of these famous machines, was light yet firm and powerful."*

At the same sort of time, B.H. Davis of the *Autocar* took a well-used Light Thirty to cover the Highland Six Days Trial. After 1,000 miles with the Daimler tourer, he wrote: *"Its inspiring ideal is an engine which shall suppress its own existence even when it is all-out on bottom gear, and which shall run 25,000 miles with no attention. The engine is more suggestive of steam than of petrol. In its presence one would never mention such words as 'revolution' or 'exhaust', for it does not seem to possess either. Treading on the accelerator pedal does not produce the familiar variations of sound or tremor. The sole effect is that a smooth, invisible flow of power, which might well be compressed air, eases off or intensifies its secret pressure."*

In November 1923, a new 35hp chassis was introduced for the 1924 season, with four wheel braking (at last becoming a standard feature across the entire range for 1925) and a lighter frame. An important change was also incorporated in the engine. Until now, the valve sleeves had always been made of cast iron, but it was decided to use light steel sleeves instead. This resulted in the lightening of the moving parts, which gave far better performance. In addition, it was found that the walls of the steel sleeves could be machined thinner, thus increasing the cylinder bore, which in turn produced more power. The 35hp engine displaced 5,764cc, although it had the same bore and stroke as the old 30hp, which displaced only 4,962cc. This anomaly was caused by the engine dimensions and sleeve valve power-units. The capacity of sleeve valve engines would be more informative if measured on the sleeve bore, not the cylinder bore. At first, these steel sleeves were only employed on the 35hp, but by the middle of 1925, they had been adopted on all Daimler engines. The 35hp chassis cost £900, while a tourer was put at £1,075, a saloon £1,125, and a De Luxe version of the closed car, £1,300.

King George's fleet of high powered Daimlers, which had covered many thousands of miles during the war period, were replaced by four new up-to-date vehicles; two Hooper 57hp Shooting Brakes, and two Hooper 57hp limousines. These 9.4 litre 57hp vehicles were no longer listed, but the Company decided to produce a number of them to fill His Majesty's order, which was being managed by Stratton-Instone. Those left over were bought by Stratton-Instone, and disposed of through the usual trade channels. All four of the Royal cars were finished in Royal claret, and delivered during the August and September of 1924, along with a smaller 20 hp limousine.

At the start of 1925, there were four 45hp models, four 30hp models, three 20hp, 21hp, 25hp and 35hp models, and two 16hp cars, the chassis ranging in price from £425 to £1,400. At this time of course, Daimler, as a producer of luxury motor cars, found itself well protected by the McKenna duties imposed on foreign cars, so larger models always seemed cheaper in relation to similar French, German or Italian marques. It will be noted that the Daimler radiator was slightly higher than it had been in the previous year.

A new series of long stroke six cylinder sleeve valve engines was introduced in July 1925, all featuring steel sleeves with larger rectangular ports and aluminium alloy pistons. Dual ignition was also standard, and full

Below: *A 1922 design proposal for a de luxe landaulette suitable for the forthcoming 16hp model.*

Left:
A single cabriolet by Hooper on the 1923 21hp Daimler chassis, body 5721 on frame number 19567.

Right:
A Hooper bodied 1924 35hp Daimler. The car shown here sports four wheel brakes, soon to become standard across the range.

pressure lubrication appeared in all but the N.30 model which retained its dated trough system. The new engines were designated as the 16/55, the 20/70, 25/85 and 35/120. The steel sleeves gave the Daimlers a new lease of life; for example, the 25/85 is said to have had an equal performance to the old 35hp car, which itself was no slouch, even when compared to some of the contemporary Sunbeams and Vauxhalls. All of these late 1925 cars had a dropped front axle, front brakes, shock-absorbers all round, four speed gearboxes (except the 16/55 which had a three speed) and a deeper radiator which gave a near horizontal bonnet line.

The Double-Six

"At a time when most manufacturers are engaged in the process of producing ever smaller motor cars, it is very gratifying to find a British firm of high reputation with sufficient faith in really big cars to offer an entirely new model of such impressive proportions as the interesting Daimler Double-Six."

1926 will go down as being a very important year in the history of the Daimler Motor Company. It had been established for thirty years, and during this time, motor vehicles had developed beyond all recognition; Daimler had never been a company to lag behind, or

Above: *A 45hp interior drive Daimler limousine by Vanden Plas, delivered in July 1925.*

shown itself to be content to follow the footsteps of others. The time had now arrived for the firm to try and produce a vehicle that could justly be termed as an engineering masterpiece - a vehicle which would confirm the Company's position at the top of the British motor industry. The call for multi-cylinder engines was rapidly expanding. For example, of the cars exhibited at the 1925 Olympia Motor Show, about 65% were four-cylinder cars, with just 27% six cylinder. By 1928, however, about 45% had six cylinder engines.

For many years, the Daimler Company had been building six cylinder engines almost exclusively, and when it was decided to design a new luxury high speed car, the number of its cylinders was a matter of supreme importance. It was intended that the power-unit should attain a degree of quietness and absence of vibration never previously witnessed with the internal combustion engine. A double six, or twelve cylinder unit, as used by Packard, Voisin and Hispano-Suiza, was duly selected for the new car, with two blocks of six cylinders arranged in a 60 degree V operating on a common crankshaft. Each 25/85 based block was to have its own carburettor and ignition system.

To say that this car caused a sensation when it made its debut in October 1926 is something of an understatement. Its twelve cylinders of 81.5mm x 144mm bore and stroke, gave the engine a capacity of 7,136cc. Dual ignition was provided, with a separate magneto for each block of cylinders. The cars track dimension was measured at 4ft 9ins., and while the wheelbase was put at 13ft, the overall length of the vehicle was a massive 17ft 8ins. The chassis sold for £1,850 and the standard saloon body could be bought for £600; it was in a class of its own.

Laurence Pomeroy (Daimler's Chief Engineer at the time, having joined the Company in 1925) could be rightfully proud of his creation, which was Britain's first ever V12 engine designed purely for automotive use, and certainly one of the finest ever made. During the earlier part of 1926, the Company was still marketing the old R and N type 30hp and 45hp cars, although by now, they were at the end of their production run. The recently-introduced 16/55, 20/70, 25/85, and 35/120 cars (now with servo-assisted braking) were carried on well into 1929, by which time, an enormous number of models were listed. In 1927 for example, 74 different variations were catalogued, including options on a lightweight version of the 20/70, said to be every bit as quick as a 4 litre Bentley.

The *Autocar* summed up Double-Six motoring after a long tour, by saying: *"There is something exceedingly fascinating about the even and effortless flow of power from the twelve cylinders, and once experienced, a strong desire for further runs such as this is aroused"* and claimed that *"quiet speed from 2mph to 82mph on top gear without the exercise of apparent effort"* was possible.

Daimler announced another smaller Double-Six in the summer of 1927, and it became available during the early part of the following year. This 31.4hp version made use of two 16/55 cylinder blocks, and had a capacity of 3,744cc. The bigger V12 was now known as the Double-Six 50, while the smaller model was christened the Double-Six 30. To distinguish these rather special machines from the six cylinder cars, a vertical line ran down the centre of the radiator; the earliest Double-Sixes did not carry this feature.

Of the smaller V12, the *Autocar* said; *"While it lacks nothing in dignity, it creates the impression that it is light to handle and manoeuvre in confined spaces, and not difficult to keep in order; hence the prospective owner-driver can examine it with enthusiasm."*

This positive move into V12 engines put Daimler streets ahead of Rolls-Royce, for although they had built V12 aero-engines during the First World War (both the Falcon and the Eagle had this configuration) it was not until the 1935 Motor Show that the famous concern introduced their equivalent, by which time, it had lost much of its appeal due to the continuing effects of the Depression.

Other News

New Directors for 1927 included Patrick Hannon (later Sir Patrick Hannon MP); Captain Edward Wood, and Arthur Pollen, the ex-BSA Chairman known for his dictatorial leadership qualities. Pollen would stay on the Daimler Board until his death in January 1937, but did not make a massive impression on future policy.

In December 1927, Minerva of Belgium had renewed their licence to produce the Knight sleeve valve engine (they had actually adopted the Knight unit before Panhard & Levassor), but the Willys-Overland-Crossley concern was also named in an agreement. Previously, Willys-Overland had built under Knight's patents in America (since 1914), but the British cars produced by the Crossley subsidiary had needed Daimler's permission.

A major shake up of distributors occurred in mid-1928, Rootes Ltd loosing their London dealership, along with nine others in the UK. Stratton-Instone were given very favourable terms, even controlling a

Below: *This Daimler limousine was exhibited at the 1925 Olympia Motor Show by Hamshaws, who bodied the car.*

Right:
This 1925 R30 landaulette by Hooper was built for HM the Queen of Spain.

Below:
A 1926 20/70 Daimler limousine bodied in aluminium by Charlesworth of Coventry for the Alifin of Oyo.

number of the other eleven dealers named by Daimler. However, this new arrangement was to be a short lived one.

The BSA factory in Coventry Road, Birmingham, had been placed in the hands of Daimler. They sold it to Singer in 1927 for £210,000, and nearly 90,000 shares in the Company. These were later sold off, producing a good profit by the end of 1929, as at this time, the Singer concern was still thriving. Had the Company have retained them longer, they would almost certainly have lost a substantial amount, as by 1934, Singer shares had dropped to around 60% of their former value.

More Developments

In May 1929, a new aluminium piston design at last reduced oil consumption on the sleeve valve engines down to a more acceptable figure. It was reported at the time that 43% less oil was used, and the Company recommended customers to convert to aluminium pistons as soon as possible. Not long after, in July, another Hooper limousine was delivered to the Royal Mews. Built on the 25/85 chassis, the body was not altogether dissimilar to that supplied on the Double-Six 30hp chassis to the Queen in the previous year. The coachwork was grey with black wings and valances and looked particularly handsome - the body number of this car was 7208. A Hooper-bodied Brougham was also delivered at about the same time. This car was based on a Double-Six 30 chassis and carried the body number 7209, finished in the traditional Royal colours.

A new six-cylinder 25hp owner-driver car of 3.57litres (later also known as the 20/30 when equipped with the fluid-flywheel) was developed in time for the

Above: *An unusual Compton boat tailed body on a 16/55 or 20/70 from around 1927.*

Opposite: *This splendid machine, dating from 1926, once belonged to the Maharajah of Rewa. The panels were of stretched German silver and it weighed nearly six tons. The car has recently been valued in excess of £900,000.*

Above: *A Double-Six 50. This P type dating from 1927, features standard coachwork by Daimler and was priced at £2,800.*

Below: *The Daimler stand at the 1927 Olympia Motor Show.*

1930 trading season, but good though the Pomeroy design was, the Double-Sixes still attracted most of the attention. Two low chassis Double-Six sports cars were built; one by Thomson & Taylor to a Reid Railton design, and the other by Corsica. There was also a low-built saloon built by Martin Walter, although the chassis itself was not lowered for this design. The 20/70 was given the new designation of the 20hp, while the 35/120 became the 35hp and the 16/55 was dropped from the range altogether.

Prices for 1930 were as follows;

20 hp chassis	*£575*
Saloons from	*£695*
25 hp chassis	*£750*
Short chassis	*£700*
35 hp chassis	*£975*
Long chassis	*£1,050*
DD6 30 chassis	*£1,050*
Short chassis	*£1,000*
DD6 50 chassis	*£1,950*
Saloons from	*£2,400*
Landau's from	*£2,600*

Compared to the competition, the smaller vehicles in the range had to prove themselves against an endless list, but the larger cars had to face the modestly priced 4 litre Bentley, the Lanchester Straight-Eight chassis at £1,325, and the new Rolls-Royce Phantom II chassis at £1,900. Foreign competition was not so fierce, with Mercedes-Benz pricing their six-cylinder 38/250 chassis at £2,150.

The Fluid Flywheel

Throughout its long and eventful history, a characteristic of the Daimler Company had been its readiness to investigate new and unusual inventions with an open mind, in spite of the opinion of others, the sleeve-valve engine had been an excellent example of this. In 1930, the Company decided it was going to abandon the use of a conventional clutch and adopt a flexible hydraulic drive, commonly known as the fluid-flywheel. They also discarded the so-called "crash" gearbox - Emile Levassor had summed up its operation in the early 1890s by saying, *"It's brutal, but it works!"*

The fluid flywheel was an adaptation of an old engineering principle, introduced into the world of the motor car by Laurence Pomeroy (Pomeroy, who had been the Company's Chief Engineer for some time, was finally appointed Managing Director following Percy Martin's resignation in 1929), who made the most of the 1926 patents of Mr H. Sinclair.

It was basically a scaled-down version of

Below: *Another 50hp Double-Six. This fine example being the 1927 Show car exhibited by Mulliners of Northampton.*

Above: *Royal Daimlers. l to r: The First Royal car after its first modification; Queen Mary's 1913 Hooper 45hp Special limousine (chassis TC 11777); King George V's 1929 Double-Six 30 and a 57hp Shooting brake, one of two such machines ordered in 1924.*

Below: *A 1929, 20/70 Daimler with body by Weymann.*

Above: *This Double-Six 30 took part in the Monte Carlo Rally in 1930. The crew consisted of S.C.H. "Sammy" Davies, Rupert Instone (son of Ernest Instone), L.C. Sharpe, A.G. Throssell and G. Tomlinson. The Daimler team completed the course but failed to win any trophies.*

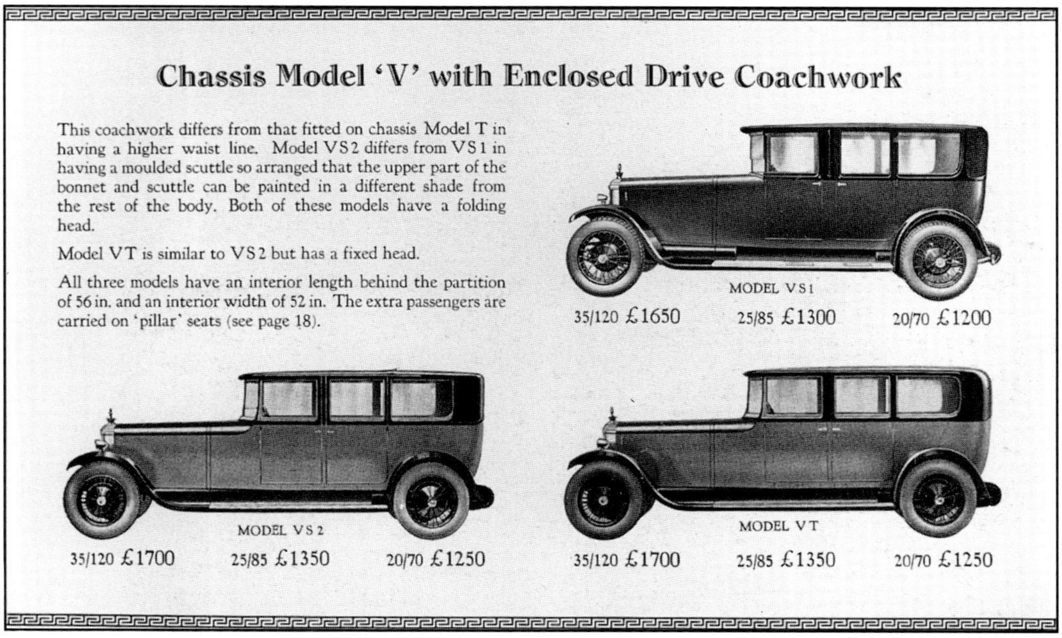

Chassis Model 'V' with Enclosed Drive Coachwork

This coachwork differs from that fitted on chassis Model T in having a higher waist line. Model VS 2 differs from VS 1 in having a moulded scuttle so arranged that the upper part of the bonnet and scuttle can be painted in a different shade from the rest of the body. Both of these models have a folding head.

Model VT is similar to VS 2 but has a fixed head.

All three models have an interior length behind the partition of 56 in. and an interior width of 52 in. The extra passengers are carried on 'pillar' seats (see page 18).

Above: *The first of the impressive low chassis Double-Sixes. This 1930 50hp model was built for C.B. Wilson under the direction of Reid Railton at Thomson & Taylors*

Below: *The Fluid Flywheel was a revolutionary form of power transmission.*

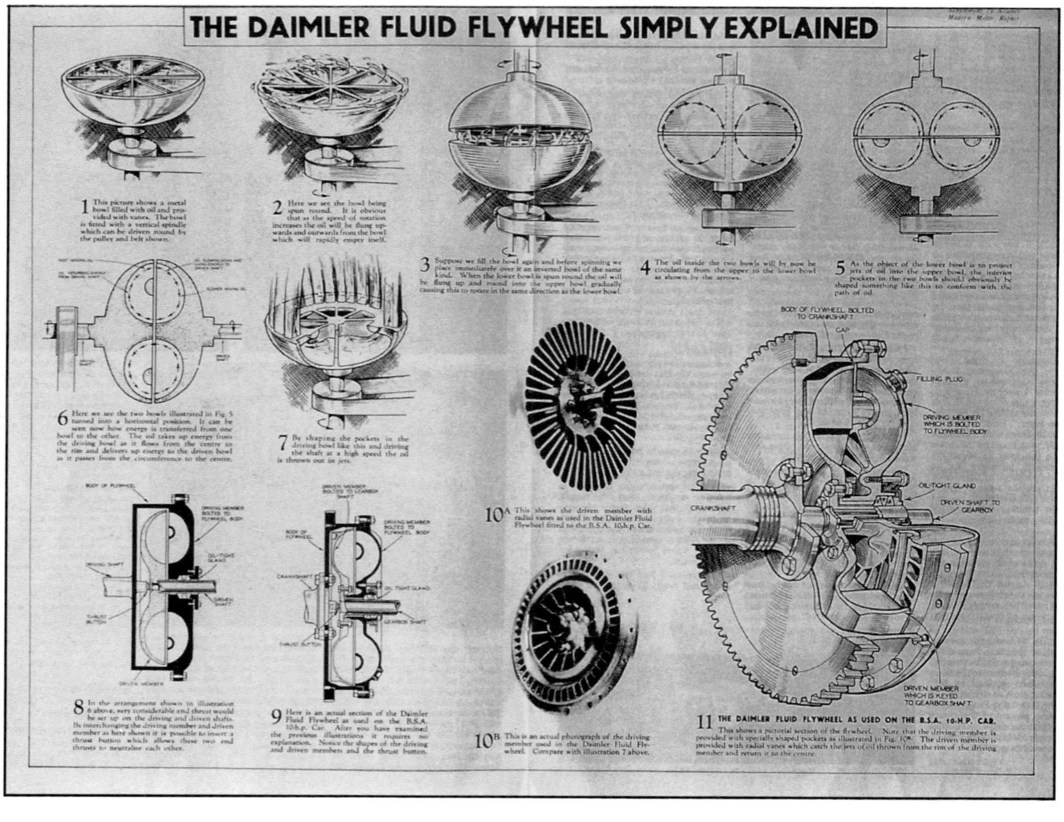

Top: *A 1930 Daimler 25/85 with Mulliner Magic Carpet coachwork.*

Bottom: *A Weymann sports coupé body on a 1929, Double-Six 30 chassis.*

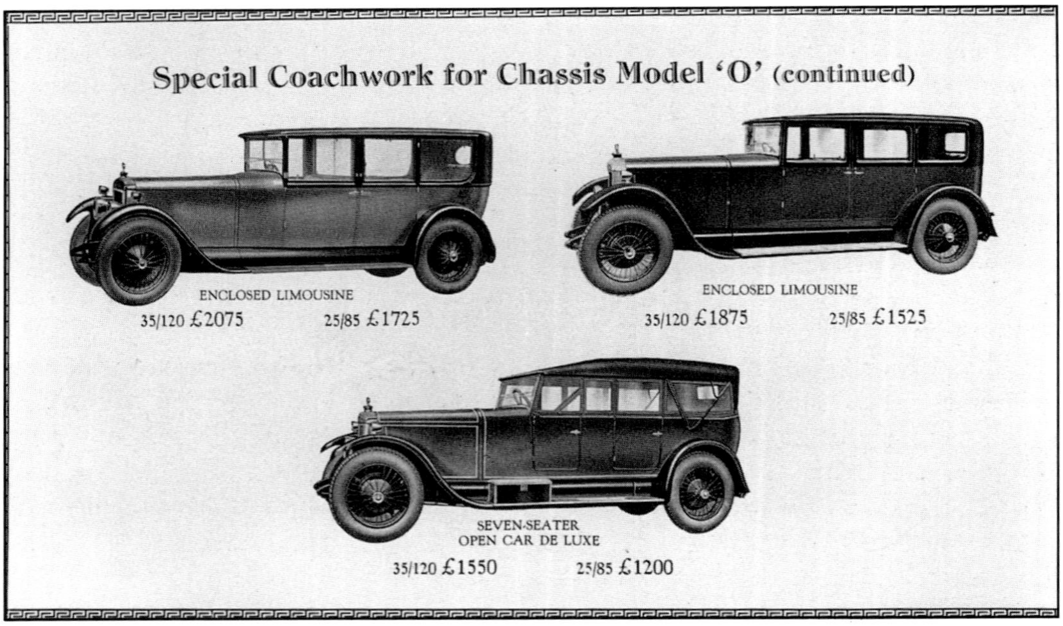

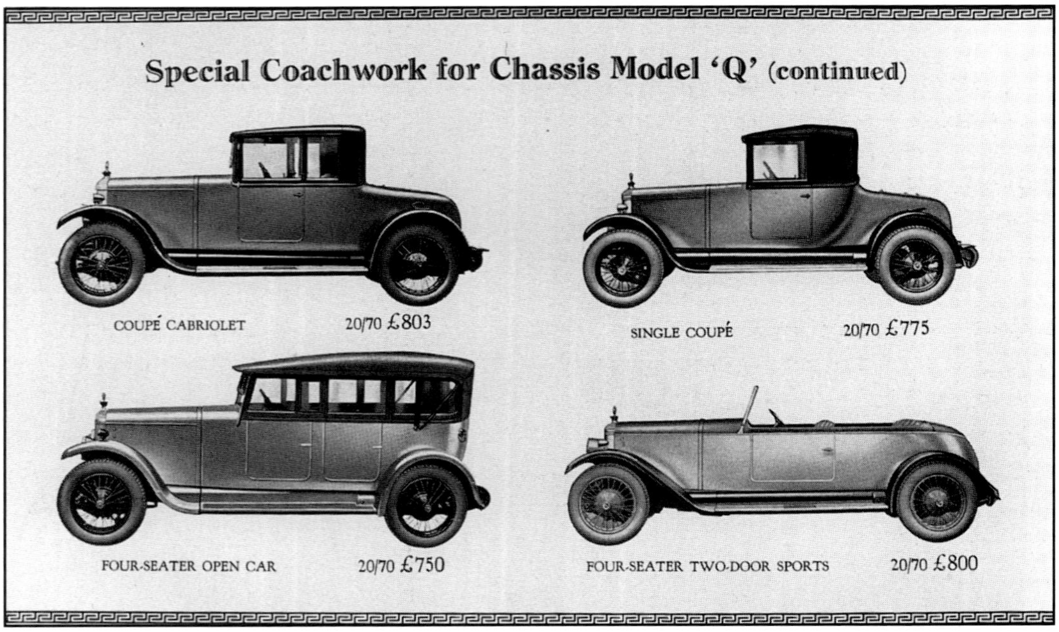

the German Vulcan-Föttinger coupling, already used fairly extensively on diesel-engined ships, but totally new to the motoring public.

The flexible drive was originally intended for use with an ordinary cone clutch and a conventional gearbox. The effect of this combination meant that it was unnecessary to declutch when the car was stationary, but it was largely due to the encouragement of Dr Lanchester and Percy Martin that experiments were made with a Wilson epicyclic pre-selector gear, as used by Armstrong-Siddeley. It soon became apparent that in this combination, a striking advance in power transmission was to be found. This advance, linked with a gear that was not only foolproof, but practically unbreakable by even the very worst of drivers, resulted in Daimler adopting the system for all of its cars in 1931. Needless to say, this new form of transmission brought about a certain amount of armchair criticism. Pomeroy was told that it would be impossible to design an oil tight joint, and that there would be *"considerable slip"* between the driving and the driven member, resulting in the loss of power and a corresponding increase in fuel consumption. Electric ignition, the pneumatic tyre, and even four wheel brakes had all undergone similar condemnation based on theory, when they were first introduced.

Eventually, the system was sufficiently refined to be put into production, and it continued to be a feature of the Daimler range right up until the mid-1950s. All critics were silenced, however, in August 1933, when the RAC carried out an independent test. An 18hp Lanchester was fitted with a device that could lock up the flywheel, thus allowing the driver to have it fluid or locked. In the hill-climbing test, the fluid-flywheel performed 15% better, and fuel consumption overall was less than 1 mpg worse. Its only real drawback was that it did not allow push-starting, as the fluid flywheel can only transmit power one way - from the engine to the wheels.

Daimler were destined to be the only real supporters of the fluid flywheel, even after it had proved its worth. One man, however, Mr T. Watson Paterson of Liverpool, did try to introduce a similar system in the mid-1930s, but nothing more was heard of the Paterson Fluid Coupling

Daimler Commercials

As far as the Daimler commercial vehicles were concerned, a period of relative inactivity followed the war, passenger models being virtually the same as their pre-war counterparts, except for the addition of electric lighting and pneumatic tyres. During 1926, the Daimler Motor Company amalgamated with the Associated Equipment Company (AEC - which had previously been the London General Omnibus Co.), with whom they had had a close working relationship since before the war.

A new company which embraced the original interests of both parties was formed. Its title was the Associated Daimler Company, and Laurence Pomeroy was appointed as Chief Engineer. The Associated Daimler lasted for just three years, and then it was decided to run the two businesses separately as before. However, several ADC series vehicles were built, and the name of the Associated Daimler Company continued to be owned by Daimler as a non-trading concern.

In the meantime, the first of the post war designs were beginning to find their way onto the market. The lightweight CM made its debut in 1926, to be followed shortly afterwards by the extremely popular Daimler CF6. This vehicle was available with either forward or normal control position, the former having a one man half-cab. The CF6 was powered by a 5.7 litre sleeve valve petrol engine and featured four wheel braking, dual coil ignition, an underslung worm rear axle and the classic fluted grille, previously reserved for use only on the Company's passenger cars.

The sleeve valve engined CG6 passenger chassis with an offset conventional transmission, and the CH6 with its new fluid-flywheel and pre-selector gearbox, appeared in late-1930, but within two years, conventional transmissions had been dropped completely, and sleeve valve engines were gradually giving way to the more popular poppet valve units.

Chapter Seven

THE DAIMLER-LANCHESTER AMALGAMATION

Part of a proof for a Press Release from Walter Judd (freelance Publicity Agents) on behalf of Daimler and dated 1st January 1931, read as follows:

"The BSA Co. Ltd is now-a-days what is known as a Parent Company or Holding Company, and therefore the transaction which has just been concluded does not imply any particularly important changes, nor any amalgamation in the ordinary sense of that term. The Lanchester Motor Company in its new status now becomes a member of that group of Companies which are under the aegis of the Birmingham Small Arms Co. Limited.

"The acquisition of the Lanchester Motor Co. by the BSA Co. Ltd is in the nature of an act of industrial rationalisation, inasmuch as the general character and reputation of the Lanchester car is in keeping with that of the Daimler Co. which is the other motor car manufacturing concern in this group. A wider and more vigorous selling policy will no doubt be possible, and it is in fact already announced that Stratton-Instone Limited will be sole distributors for Lanchester cars, and that Mr Frank Lanchester, who has been the London Director of the Lanchester Motor Co., will join the Stratton-Instone Board".

Facing Page: *A Rippon bodied touring limousine for owner or chauffeur on the 1938 32PS (4.5 litre) Straight-Eight Daimler chassis.*

By 1931, the Double-Six 50 had been dropped from the Daimler range, and two new twelve cylinder cars were introduced to complement the original Double-Six 30, which stayed in production until 1932. Neither of these first V12 models had been fitted with fluid-flywheel transmission, although it was a standard feature on the Double-Six 30/40s and 40/50s, which had made their debut in September 1930. These new models, the Double-Six 30/40 with its 5.3litre sleeve valve engine and slightly taller and more slender radiator shell, and the 6.5litre Double-Six 40/50, shared many of the features that had been introduced on the six cylinder 25hp Daimler.

With the new fluid flywheel, the massive and much modified engines of the Double-Sixes seemed even smoother and more flexible than before. With aluminium heads now, as well as the monobloc crankcases, these were probably the first all alloy V12 engines in the world.

The smaller six cylinder models were kept much the same as they had been in 1930, with the single exception of the 20hp car which was given a new "16-20" designation from March 1931. The wheelbase on this vehicle was put at 10ft 1ins. in 1931, but just over a year later, it was extended to 10ft 4ins. and incorporated several body and chassis modifications, including a new radiator with a centre strip (as seen on the Double-Sixes) and what can only be termed as shoulder ledges. For a time, some of the smaller Daimler cars were offered with a conventional sliding gear transmission as an option, but the fluid flywheel system soon took over completely.

Five new Royal Daimlers were delivered during April 1931. King George V ordered three Hooper limousines - one on the Double-Six 30/40 chassis, and the other two on the Double-Six 40/50 type chassis. Queen Mary had two limousines; these were a 30/40 (registration number XE7432) painted in green and black, and a 40/50 (registration number A3179) finished in the traditional Royal colours, as were the King's cars.

The Lanchester Link

Apart from the merger between Rolls-Royce and Bentley in 1931 (a sign of the times), another major amalgamation took place, this time within the BSA Group. The Lanchester Motor Co. Ltd was taken over by Daimler, and it was in some ways fitting that these two famous marques should merge. The Lanchester car, designed by Dr Fred Lanchester, LLD, FRS, MICE, MIME, MIAE, had been this country's first ever petrol driven four wheeler.

Fred Lanchester (born in Birmingham in 1868) was one of four brothers, three of whom were involved with the Lanchester cars. After studying at the Royal

Left:
A Daimler Double-Six 50 Specially bodied by Mulliner for a Canadian enthusiast.

Right:
The lowered chassis Double-Six sports car by Corsica of London, seen here with its cousin from the 1970's.

Above: *Although it looks as if the chassis has been lowered on this superb machine, it is actually quite standard, only the Martin walter coachwork is lower than usual. It was delivered to Mr Webber of Wandsworth in 1932*

Right:
The first Lanchester, 1895.

Above: *An unusual Double-Six, delivered to Mr. Sakrealleades in the latter half of 1931.*

Below: *Despite the original caption, the vehicle in the foreground is the first Royal car of 1900, whilst the large limousine behind is one of the Double-Six 50's delivered in 1931.*

35 Years of Progress in Motor Car Design.

A 6-H.P. DAIMLER BUILT IN 1899 FOR H.M. KING EDWARD VII AND ONE OF THE LATEST 50-H.P. DAIMLER CARS WITH "HOOPER" BODIES BUILT SPECIALLY FOR HIS MAJESTY KING GEORGE V.

College of Science, he was appointed Assistant Works Manager for T.B. Barker & Co. of Birmingham - the makers of the Forward gas engine. During the early 1890s, Lanchester began to design his first car in a small workshop rented from his employers.

He made a study of every Continental vehicle, but was unimpressed by any of the existing designs. It is said that he regarded the Panhard, which was perhaps the leading French car of that period, as a very poor specimen when judged from a purely engineering point of view. In his opinion, there was no excuse for the exposed gears then used, nor for the lack of any attempt to try and balance the engine. To him, everything he examined served more as a warning than as a guide.

The first Lanchester car of 1895, had a chassis frame consisting of nine weldless tubular transverse members of 2ins. diameter with plate sides. The original engine was a balanced 5hp single cylinder unit, though this was later changed to an air-cooled 8hp twin cylinder engine, fitted with two flywheels which rotated in opposite directions to help balance it. Both the engine and transmission were mounted on an undercarriage, the original chain drive having being abandoned in favour of a worm drive cut on a special machine of Lanchester's own design. Lanchester was actually one of the first men to adopt the worm drive, and for many years, it was a feature of all Daimler products.

The second car designed and built by Lanchester appeared in 1897, and it was a considerable advance over the first. The transversely mounted twin-cylinder engine was air-cooled, and each piston was coupled to a separate crankshaft. These crankshafts, one above the other, were connected by skew gearing and ran in opposite directions in order to obtain engine balance. A flywheel was carried on each of the crankshafts, but at opposite ends to each other. The ignition system was very modern for its time; one of the flywheels carried two permanent magnets which revolved around a fixed armature, and an ingenious system was adopted to prevent fouling at the points. This vehicle was awarded a Gold Medal at the trials organised by what was then the Automobile Club of Great Britain and Ireland (now the RAC) at Richmond in Surrey, during the summer of 1899, and it also took part in the famous 1000 Miles Trial of the following year.

In 1897, premises were acquired in Ladywood Road, Birmingham, and Lanchester was joined by his brother, George (b. 1874). When the Lanchester Engine Company was formed in the December of 1899, Charles Vernon Pugh was appointed as the Company's Chairman, with F.W. Lanchester himself taking the position of Manager and Frank (b. 1870), another of the brothers, was named as the Secretary. The business started in the Armourer Mills, Montgomery Street, Birmingham, and work began at once on a new car. This air-cooled 10hp model, placed on the market in 1901, was based on the two original designs, but was even more advanced, and was followed by a series of higher-powered cars, up to the 20hp of 1904, capable of 50mph.

Above:
Dr. F.W. Lanchester - one of the country's finest engineers.

Vertical engines were becoming an almost universal feature of new cars, and it was abundantly clear that any manufacturer who persisted in using a horizontal engine would be fighting an uphill battle. So, in 1905, the Lanchester Motor Co. adopted a vertical engine for their new model. The car was completely redesigned, with the four cylinder power unit being rated at 20hp. It was later joined by a six cylinder 28hp, which was increased in 1910 to 38hp. In the following year, a four cylinder 25hp vehicle was introduced - the last Lanchester car to be designed by Fred Lanchester.

Future designs were the work of George

Left:
A 12hp water cooled Lanchester.

Left:
A pre-1914 Lanchester car outside the Company's impressive London depot and showrooms at 95 New Bond Street.

Lanchester, and for a number of years, the Lanchester car continued to display a degree of individuality shared by very few of the other marques. Despite their differences, however, there has always been a certain affinity between the Daimler and Lanchester cars. Both were pioneers and both played an important role in the birth of the British motor industry.

The Sporting Forty made its debut just before the outbreak of the First World War, and formed the basis for the superb Lanchester Forty, built between 1919 and 1929. Around 500 of these luxury 6.2litre six cylinder ohc machines were built, several going to Royal customers. A smaller version, also marketed as the Forty, was a very expensive motor car, but both were to be overshadowed by the 192830 hp Straight-Eight. Unfortunately, this 4.5litre 82bhp carriage, which inspired Pomeroy to build his straight-eight, came at just the wrong time and sales were very low due to the Depression. The last few were built towards the end of 1931, the very last one being sold in 1933.

In January 1931, the bank decided to call in their overdraft (some have said BSA encouraged them to do so), forcing the Lanchester Directors to either amalgamate or close the business completely. For just

Left:
Straight-Eight Lanchesters were extremely fast and refined cars in their day. Production of these continued for a while after the Daimler lake over, despite their high price.

Below:
A Lanchester Forty landaulette from the mid 1920s.

Left:
Colonel A.H. Loughborough with the Lanchester 15/18 with which he won the first ever RAC Rally, the event was held in March 1932. Incidentally, a Daimler finished in second place, driven by J. Mercer and a certain "Lofty" England of Jaguar fame.

Right:
A Daimler 25hp limousine specially built for the 1931 Buenos Aires Exhibition. Note the beautiful Lalique radiator mascot.

£26,000 they merged with Daimler who, also struggling financially, were looking to expand into the cheaper end of the market without a loss of prestige, and saw the Lanchester name as an ideal opportunity for badge-engineering. This would have been another blow for Dr Fred, as BSA had cancelled his consultancy in 1929, and due to the cost, had also closed Lanchester Laboratories, his research and development firm.

The Lanchester plant was transferred to the Radford works, and the Burton Griffiths Machine Tool Co. (a BSA subsidiary) moved into the empty Birmingham building. George Lanchester became a member of the senior design staff, although he was apparently never happy, and Frank was made Sales Director for the newly-formed Lanchester division.

The 1931 Press Release from Walter Judd (referred to on chapter page) on behalf of Daimler also contained the following:

"It is announced that the Birmingham Small Arms Co. Limited have acquired the whole of the share capital of the Lanchester Motor Co. Limited. The negotiations for this transfer of ownership were completed shortly before the death of Mr Hamilton Barnsley which occurred during the Christmas holidays. Mr Hamilton Barnsley was Chairman, Managing Director, and also principal shareholder in the Lanchester Motor Co. For some time past he had not enjoyed the best of health, although his death was both unexpected and premature, and he opened negotiations with the BSA Co. primarily with the object of relieving himself of his responsible duties.

Above:
A 1932 Daimler 16/20 saloon.

Below: *Percy Martin.*

"The Lanchester Motor Co., as is well known, is one of the pioneer firms in the motor trade. In fact it may be said to have been the British pioneer firm on account of the strikingly original design of its early products which were developed on lines that were quite different to the general trend of Continental design which dominated early practice.

"It was natural for Mr Hamilton Barnsley to approach the BSA Co. for they are his immediate neighbours, inasmuch as the factory premises of the Lanchester Motor Company adjoin those of one of the BSA factories.............

"Owners of Lanchester cars will in future have the benefit of a more extensive service organisation, and Mr George Lanchester who designed the cars and was the Director in charge of the works, will continue to occupy a similarly responsible position."

It was going to be interesting to see how Daimler were going to service these unique cars, having scrapped most of the spares and sacked all but three of the firms employees, but as soon as the merger had been completed, despite assurances that *"in no sense will there be a fusion of the Daimler and Lanchester cars"*, a new 15/18hp six cylinder Lanchester was introduced. A Daimler fluid flywheel was fitted, and a feature of this model, which made its debut late in 1931 for the following season, was its 2,504cc ohv engine. This was the first move by the Daimler Motor Company to abandon the use of the sleeve-valve engine, and to revert to the poppet valve. There were a number of good reasons behind this decision.

When the sleeve valve power unit was first adopted in 1908, metal science, valve design, engine balance and silence of running, together with the many side-issues, were still at a relatively early stage of development. The poppet valved engine of 1908 was both rough and noisy, but the sleeve valve engine had overcome these drawbacks in one stroke. The undeniable advantages of the sleeve valve caused advocates of the poppet valve to experiment, and devise ways and means of improving it.

The First World War had effectively advanced internal combustion engineering, and all the lessons learned during the conflict were instantly put to use as soon as the motor industry got back on its feet. Valve lifting cam contours and tappet design were at last taken seriously, and as engine speeds increased and silence became a demand made by the car buyer, both underwent a great deal of development and refinement.

As a result, the poppet valve engine of the 1930s was a far cry from its predecessor of twenty-five years earlier, and the wide gap which had separated the sleeve valve from the poppet valve was, in the course of time, almost bridged. As poppet valve engines were far cheaper to produce, the time had come to gradually withdraw the sleeve valve design from production. There is little doubt that Daimler would have replaced the design earlier, but a lack of funding from BSA prevented the Company from developing and introducing their new power-units any sooner.

The *Autocar,* in their test of the 15/18 Lanchester, noted *"its running on the road is exceptionally good and exceptionally pleasant."* They added, *"Possibly the outstanding feature of the car is, paradoxically, that it has no outstanding feature, since the design is so well-balanced as to make it universally good throughout."* This would have undoubtedly calmed George Lanchester down a little, as he had been pushing Daimler to fit an ohc engine.

The 15/18 was very much a compromise design, bridging the original Lanchesters and the new Daimler-Lanchesters.

Other News

The very first RAC Rally was held in March 1932, and was won by Colonel A.H. Loughborough in a Lanchester 15/18, while second place went to J. Mercer in a Windovers-bodied 30/40 Daimler, co-driven by "Lofty" England, a Daimler apprentice, and later to be the Chairman of Jaguar. The Daimler range for 1932 consisted of six models; the 16/20, the 20/25, the 35hp, the Double-Six 30, the Double-Six 30/40, and the largest of them all, the Double-Six 40/50.

1932 witnessed a partial recovery from the serious financial crisis that Britain had faced during the Depression, the Company at last seeing a profit again after being in the red since just before the Wall Street Crash, although dividends for the shareholders were still a long way off. Small, high class cars which could be run economically were the order of the day, and in response to this new demand, the Company brought out an entirely new car in October 1932. The 9.9hp car was badged as a Lanchester, and was a full four seater with a fluid flywheel and preselector gearbox. The car had a ready market, for its purchase price was very

Below: *Sir Edward Manville.*

Above: *The Daimler Fifteen was first introduced in the 1933 model year range. This 1935 example carries a six-light saloon body by Mulliners.*

reasonable for the class of vehicle, and its running costs were low. Various coachwork options on a fairly low-built chassis even allowed *MotorSport* to include the open four seater (£350) in their "Sports Cars for 1933" list.

In addition, a new line of BSA cars was based on the 1933 Lanchester models, employing side valve engines and fluid flywheel transmission, far better than their 1.1litre air-cooled V-twins bought from Hotchkiss in the 1920s. A new range of Daimler-Lanchester distributors was established, taking away the privileged arrangement with Stratton-Instone at the same time, an arrangement that had seen them as sole distributors for London and the adjoining counties since August 1928. BSA were highly critical of the Stratton-Instone agreement, and rightfully so, as the dealers were on a very high commission rate. The BSA Director's accused the Daimler Board of *"living in the past"*, and not applying modern practices in line with the growth of the business, though it has to be said, that this was not the case generally.

Since the retirement of Sir Edward Jenkinson in 1906, the Chairmanship of Daimler had been held by Sir Edward Manville. Manville was a trained engineer, who had been associated with motor cars almost from their birth. In 1900, he managed to purchase one of the 6hp Daimlers that had put up such a remarkable performance in the Thousand Miles Trial of that year. He was an active member of the RAC, serving on various committees and, in 1907, he accepted the office of President of the Society of Motor Manufacturers & Traders - the highest honour that the motor industry could bestow on an individual. Financed by Dudley Docker, Sir Edward also tried his hand in politics, and was to remain the Chairman of Daimler until his death in March 1933 - a period of some 27 years.

Shocked by this great loss, ironically it was Sir Alexander Roger (one of the first business candidates for Parliament, and BSA Chairman from 1906 until 1928, when Manville took his place) who briefly took the Daimler Chair. Roger promised much but delivered little before Percy Martin was brought back to take up this important post, in January 1934. However, Sir Geoffrey Burton (who joined the Daimler Board in April 1934), was elected as Chairman in December 1936 following the retirement of Martin - a post which he was to hold for the remainder of the pre-war period. This was the end of Percy Martin's involvement with the Company, but he had served it well, forcing it into the modern machine tool age. After his official retirement, he devoted his time to breeding pedigree cattle on his Kenilworth farm. He died in 1958.

The Daimler Fifteen

At the same time as the Lanchester Ten, the first poppet valve Daimler since 1908 was introduced. This came in the form of a medium-sized 15 hp six cylinder model, and was again aimed at that section of the market that wanted a thoroughbred car which was easy

Right:
A Daimler Light Fifteen.

Right:
A delightful little 10hp BSA saloon (model 102) dating from the late summer of 1934.

to handle, reasonably inexpensive to run, and did not require a chauffeur to look after it.

This 15hp model, was one of the most refined British cars of medium size built during the pre-war period. It handled well, and its overhead poppet valve engine was extremely quiet. The car went on to sell in great numbers, and its popularity was just as apparent in 1939 as it was when it was first introduced. There are some who consider the Fifteen indirectly responsible for the gradual demise of the Daimler marque, as some of the more well-to-do Daimler owners were aggrieved to find lesser mortals with Daimler cars. This may or may not be true, but the Company would have certainly struggled through this lean period without it. The Press loved the car, however, the *Autocar* praising virtually every aspect of the vehicle before coming to the conclusion that *"the Daimler Fifteen is undoubtedly a remarkably attractive proposition."*

It is interesting to note that Rolls-Royce were also considering producing a car of a similar size. They never did, and in fact had reservations about introducing their highly-successful 20hp model at first, without which they too would have been in trouble. Other Daimler vehicles included in the range for 1933, were a 20hp type LQ chassis (previously the 16/20), a 25hp (the old 20/25), a 40hp Double-Six, and a 50hp version (once the 30/40, and 40/50 models respectively) - all of which were fitted with sleeve valve engines.

Right:
A Light Twenty with a standard four light saloon body, circa 1936.

Below: *A Salmons-Tickford cabriolet body on the 1935 Daimler Fifteen chassis. This particular car now resides in Holland.*

Above: *This impressive Daimler, registered in July 1935, had a rather short but eventful life. It was originally built as a Double Six, but was fitted with a straight eight engine by the Experimental Department at the factory. By the following January the car had been scrapped.*

Below: *A Drophead Coupé on the Daimler Twenty chassis.*

The 1934 Range

The 26hp, 40hp. and 50hp were left much the same for 1934, but the 20hp model was given a new poppet-valve 2.7litre engine in September 1933, and this latest Daimler was designated Type LQ2.20. In limousine form, the chassis was extended 3 inches to give a wheelbase of 10ft 4ins. From the early part of 1934, this limousine was given the new Type reference LQ3.20. In this guise, the Twenty continued in production until April 1936.

Prices were as follows;

Lanchester Ten

Chassis	£245
Saloon	£325
Sports saloon	£345
Sports coupé	£335
Holbrook saloon	£357
Denton coupé	£380
Romney coupé	£380
Tickford coupé	£385

Left:
Hoopers built this Lanchester limousine for HRH the Duke of York (later King George VI) in 1936. It used the Daimler 4 ½ litre Straight-Eight chassis, and was one of a pair of Lanchesters delivered in February 1936.

Right:
Queen Mary's 1935 OPR3-50 Double-Six (one of only nine such chassis types built) outside Buckingham Palace.

Above: *A special three seater drophead coupé by James Young, built for Joseph Mackle of Stratstones on the Light Twenty chassis in 1936. Mackle's cars were always guaranteed to be unique.*

Lanchester 18hp

Chassis	£450
4-light saloon	£595
6-light saloon	£595
4-light DHC	£635
Sports saloon	£625
Continental saloon	£675
Rye 2 door cabriolet	£675
Denton coupé	£645
Romney coupé	£635

Daimler 15hp

Chassis	£350
Saloon	£450
Sports saloon	£480
Sports coupé	£465
Continental saloon	£525
Rye 2 door cabriolet	£545
Denton coupé	£495
Romney coupé	£495
Tickford coupé	£495

Daimler 20hp

Chassis	£550
Saloon	£695
Limousine	£775
Saloon de Luxe	£795
Continental saloon	£795
Continental limo.	£875
Windovers limo.	£895
Wingham 4 door cab.	£850

Daimler 25hp

Chassis		£700
Mulliner limo.	from	£1,195
Windovers limo.	from	£1,200
Hooper limo.	from	£1,310

Daimler Double-Six 40hp

Chassis		£1,150
Mulliner saloon	from	£1,700
Mulliner limo.	from	£1,900
Hooper saloon	from	£1,795
Hooper limo.	from	£1,800

Above: *A 1936 Series Five BSA Scout with a tidy two seater sports tourer body. It should be noted that no BSA cars were built after the Second World War.*

Daimler Double-Six 50hp

Chassis		£1,250
Mulliner saloon	from	£1,800
Mulliner limo.	from	£2,000
Hooper saloon	from	£1,895
Hooper limo.	from	£1,900

During 1932, yet another engineering masterpiece was manufactured by the Company, or more rightfully, their Managing Director. Inspired by L.H. Pomeroy's production of a magnificent straight-eight engine, the Daimler Board decided to build a car around it which could be incorporated into the range for the 1934 season; the new model was introduced during the early part of that year. This came at a time when the decision was made to abandon the sleeve-valve altogether, so the engine was designed with overhead poppet-valves operated by pushrods, and bored out to some 3.7 litres giving a rating of 25.7hp. This really marked the end for the sleeve-valve which was retained only on the Double-Six 40hp and 50hp cars.

The advocates of multi-cylinderism had to face many difficulties, and the straight-eight engine had always possessed pitfalls for the designer. The first straight-eight seems to have been built by CGV in 1907, but they were still expensive to develop and build, and always had inherent balance and carburation problems due to the length of the engine.

The Daimler design team were fully aware of these, and the success of the Daimler Twenty-Five (or V26 as it was known in the factory), constitutes a landmark in the history of British automobile engineering. The lack of noise and vibration in the car, was little short of perfection.

Each engine was bench tested for eight hours at various speeds, then checked for any leaks, adjusted and washed down prior to fitting. In May 1934, the price of the chassis was £900, so complete, a car usually cost in the region of £1,500.

In autumn 1934, the Daimler Fifteen was slightly modified in readiness for 1935. The engine bore was kept the same at 63.5mm, but the stroke was increased from 95mm to 105mm, giving a capacity of just over 2litres. Power was thus increased to 46bhp, making the car much more lively. The model was improved

Above: *The sister car to the Royal Lanchester on the previous page, this one a landaulette, seen here with members of the Royal Family.*

Below: *The Jam Sahib of Nawanger ordered a pair of Vanden Plas bodied Daimlers built as Lanchesters on the V4 ½ litre Straight-Eight chassis. This is the tourer which was delivered in 1936. The other one of the pair was a saloon delivered in the early part of 1937. Both were finished in pale blue.*

Above:
Built to match the Jam Sahib's tourer, this Vanden Plas saloon was completed in April 1937. Other pictures of the car show the centre coachline in a much darker colour.

cosmetically too, with the radiator, headlamps and wing valances getting attention. Girling mechanical brakes with a pull-up handbrake were fitted, and the engine was mounted on rubber (as was the radiator from late 1935). Maximum speed was put at 70mph, with a fuel consumption of around 24mpg.

Apart from this, the range continued very much the same; the addition of some new coachwork options being the only substantial change. The Lanchester Ten gained some minor improvements, and four new coachwork styles (306, 307, 308 and 309). It was stated that the Ten *"is a car remarkable for its efficient overhead-valve four-cylinder engine, its Daimler fluid-flywheel transmission, and its easy riding qualities."*

On the face of it Daimler were doing well, but times were indeed hard, and the Company was struggling to make any profits at all. No sooner had manufacturers recovered from the effects of the Depression, than the Government brought in new legislation forcing up the price of motoring - horsepower taxes, and then the Road Fund licence, which even in the mid-1930s brought in an unnecessary surplus of several million pounds.

A New Era

A smaller Light Straight-Eight with a 3.4litre capacity, designed largely for the owner-driver to give "*high performance with comfort*" was announced along with a 4.6litre version for the 1936 season. In response to the higher taxes, the grandiose Double-Sixes were finally withdrawn in favour of these new eight cylinder machines - one magnificent era had passed, but another was just beginning.

The larger of the two new Straight-Eights was actually the first to appear, in December 1935. This V4½ model, rated at 31.7hp, was basically a modified version of the V26, with a slightly larger engine and higher compression ratio. The chassis dimensions remained the same, although a frame with an extra 9inches in the wheelbase (type VF) was made available from early 1937. The Light Straight-Eight (chassis Type E3½) featured a brand new engine of 72mm x 105mm, and whereas the V4½ and VF chassis were used for the more formal coachbuilt bodies, the Light Straight-Eight lent itself perfectly to more sporting coachwork. The E3½ also had a slightly livelier performance, one being timed at Brooklands at just over 90mph.

Badge-engineering may seem to be a fairly new thing, but Daimler were forced into doing just that for a number of loyal Lanchester customers. HRH the Duke of York (later King George VI), ordered two Straight-Eights badged as Lanchesters in November 1935. One of these was an enclosed landaulette by

Above: *Another Straight Eight, this time pictured alongside a contemporary aircraft.*

Below:
Based on the popular Daimler Fifteen, this Lanchester LA14-2 (or Roadrider de Luxe) was introduced in 1937 at a price of £375. It bore little resemblannce to the old Roadrider, type LA14-1, of which only a few hundred were built. Production of the LA14 (including the LA14-3 export model) ceased in 1939.

Hooper, whilst the other was a virtually standard Hooper limousine. In India, there were several notable Lanchester enthusiasts, including HH the Maharajah Jam Sahib of Nawanger. Vanden Plas built the body of his Daimler-made Lanchester for him - one of over 40 Lanchester cars that he owned.

In addition to the Straight-Eights and the (slightly improved) Fifteen, there was also the new 2.6litre Light Twenty (19.3hp) and the Standard Twenty, which by April 1936 had been uprated to 23.8hp and given the new designation "LQ3/24" - both sported six cylinder poppet valve engines. During the late summer of 1936, the LQ3/24 was renamed yet again after a number of improvements had been made to the model, particularly to the cars chassis. Its new reference was the "EL24", although it was often referred to simply as the Twenty-Four.

The Lanchester Ten, which had proved a fairly successful little car, had its engine slightly increased in size for the 1936 season. It had been found that its acceleration was rather sluggish and that its ability to negotiate hills was poor. The bore and stroke of the original engine was 63.5mm x 95mm, giving a cubic capacity of 1,203cc. For 1936, these dimensions were increased to 66mm x 105mm giving 1,444cc, which gave the vehicle the extra power it so desperately needed. Another addition to the Lanchester range was an attractive Twelve-Six, which fitted in nicely between the larger Eighteen-Six, and the small four cylinder Ten.

As well as the Lanchester Ten, there was also the BSA Ten (introduced in 1933) and for 1936, a six cylinder 12hp BSA "Light Six" was also listed. The fwd BSA Scout had made its debut in 1935 but, like the 1929 three wheeler, because this was built by BSA Cycles, it was never catalogued with the Daimlers.

By this time, George Lanchester had left the Company, ironically over an argument concerning the Tens power - or lack of it. According to Anthony Bird, *"George Lanchester became so disenchanted that he left the Daimler Company as soon as his service contract allowed."* Dr Fred had continued his experimental work, but gravely ill by the mid-1930s, died in March 1946. George moved to Alvis, and retired at the age of 87 - he died in 1970. The backroom boy of the Lanchester brothers, Frank, stayed on as the Sales Director of the new Lanchester division for a while, and died in March 1960.

Daimler Commercials

As for the Daimler commercials, the CP6, complete with a 6.6litre poppet valve petrol engine, had been on the market for quite some time, but by now diesel engines were gaining in popularity, so the Company introduced the Gardner-powered COG5 late in 1933.

Below: *A splendid 2.5 litre Daimler coupé owned by a Mr Curzon.*

Above:
This 1937 Wingham cabriolet by Martin Walter was much used by Royalty. It was built on Light Straight Eight chassis no. 40184 and is seen here during the Coronation.

Below:
This impressive 4 ½ litre Straight-Eight was built for Lady Roger, the wife of the then Daimler Chairman, Sir Alexander Roger.

Above: *Another delightful coupé.*

Left:
A 4.5 litre Straight Eight (Chassis No. 44268) with a Hooper landaulette body. Built towards the end of 1937, it featured a special hinged roof to ease access by disabled owner.

Several diesel models followed and demand for them was so great that it was decided to create another company, quite separate from the parent concern, to deal purely with commercial vehicles. A company known as Transport Vehicles (Daimler) Ltd was duly formed in 1936, and the first six-wheel Daimler trolley buses were produced not long after. By now, all Daimler buses were equipped with hydraulic brakes, and the Company had started to develop its own diesel engines.

Back to passenger cars, and for 1937 the six cylinder Daimler Fifteen had its engine capacity slightly increased from 2,003cc to 2,166cc. The famous Double-Six was revived after its year long absence, now featuring poppet valves. The V12 engine of 6,511cc had a horsepower rating of 49.4hp, and was used in a number of Royal cars before disappearing completely. The Fifteen, however, went on to be immensely successful (in Daimler terms at least), continuing in various forms until the mid-1950s.

At about this time, a new Lanchester model was introduced. The car was christened the Roadrider

This page: *Some of the Daimler Straight-Eight models available during the latter part of the period. Equivalent bodies were produced for the Light Twenty and Light Straight-Eight chassis, but whereas for example, the body style on the eight cylinder car would be no. 271, the smaller car would be type 251, and so on. For the Light Twenty chassis, the total cost of the vehicle averaged out at about £300 less expensive....*

Facing page: *The more formal 4.5 litre Straight-Eights were available with coachbuilt bodies such as these.....*

Above: *Touring limousine by James Young - £1,600*

Above: *Touring limousine by Hooper - £1,750*

Above: *Continetal saloon by Mulliners - £1,575*

Above: *Touring limousine by Freestone & Webb - £1,660*

Above: *Landaulette by Hooper - £1,685*

Below: *Six-light limousine by Barker - £1,660*

Above: *Six-light limousine by Mulliners - £1,550*

Below: *Six-light limousine by Windovers - £1,600*

Above: *This period advertisement for the four litre Daimler Straight Eight Cabriolet appeared in a late 1938 edition of Punch.*

(Type LA14) to emphasise its excellent suspension and roadholding qualities. It superseded the Twelve-Six, and was powered by a six cylinder engine of 1,527cc. With a rating of 13.4hp, the Roadrider possessed all the characteristics previously associated with the Lanchester marque.

Cyril Simpson was appointed Chief Engineer in May 1936, having joined the Company from Sunbeam in 1919. Staying on until well after retirement age, Simpson would shape the Company's future products, and see it safely into the 1960s. Another change in the staff concerned Laurence Henry Pomeroy, who failed to be re-elected to the Board in October 1936. Pomeroy, born in 1883, had gone to Vauxhall and transformed their cars overnight. He had done so much for the Daimler Company too, after joining them in 1925.

Despite their often blazing rows, Pomeroy held the greatest of respect even from Dr Lanchester, who had prompted Percy Martin to acquire Pomeroy's services in the first place. It is amazing to think that this great engineer was not re-elected, but in reply, he went to De Havilland as the General Manager of their engine division.

This great loss was not made any easier by the Company's financial situation. Daimler needed someone like Pomeroy to develop something special and give the engineers point of view in the Boardroom. Profits (when they made them) were dismal compared to sales, mainly due to the vast range offered by the Company. The Rootes brothers, who standardised the Humber and Hillman models, were recording enormous profits (Humber declared £263,000 for 1937) for their Coventry sites. The lack of strong leadership also had something to do with these poor results, as even William Lyons, with his very limited range of SS-Jaguars, managed a £34,000 profit for 1937 - something was definitely wrong at Radford.

Below: *The chassis of the Daimler Fifteen*

Further developments on the Fifteen

The 15 hp Daimler underwent several design changes during the 1937 season. Until now, an independent front suspension had failed to be adopted on the model, but a new design was introduced that provided the advantages of independent suspension with none of the drawbacks being experienced by other manufacturers at the time. This was accomplished by means of coil springs and a front axle which actually pivoted in the middle. The track and wheelbase dimensions were adjusted to 4ft 4ins. and 9ft 6ins. respectively, and although the car was now slightly heavier, the performance suffered very little.

By now, the characteristics of the post-war DB18 were becoming evident, and in 1938 the Fifteen's engine was further enlarged to 2,522cc, giving a respectable 64bhp (or an RAC rating of 18hp). In 1937, this chassis type had been referred to as the DB17/1, and then as the DB17/2 just a few months later, after suspension modifications. In 1938, however, the DB18 designation was adopted for the 2.5litre 1939 Model Year cars.

A number of variations on the DB18 theme appeared, including several delightful drophead coupes and cabriolets, as well as two special models; the Ritz and the Dolphin. The Ritz was basically a de luxe hand built saloon which sold for £575 instead of £485 for the standard version. For the extra money, the customer got a superbly finished silver-grey body with a light grey leather interior; figured grey sycamore was used for the door cappings and mouldings, and items such as a radio and picnic tables were included in the price.

The Dolphin on the other hand was an out-and-out sports car, and was entered on the 1939 Monte Carlo and RAC Rallies. The engine developed around 90 bhp thanks to twin carburettors, an aluminium head, an increase in the compression ratio, and a new inclined valve arrangement. The cars top speed was put in the region of 83mph, and acceleration was very impressive for the time.

George Fabel clearly remembers lightening the chassis of the rally car (DVC 681) by drilling a series of holes down the side members. It appeared with four-light coachwork on the Monte Carlo Rally (with Bob

Below: *Model 652: A four door sports saloon body on the Daimler Fifteen Chassis. For the 1937 trading season, this would have cost just under £525.*

Above: *Model 653: A rare four door Tickford Cabriolet on the 2.5 litre Daimler Fifteen chassis costing £645.*

Below *The Daimler Ritz - a very special saloon.*

Right:
A rare picture of the Dolphin with its saloon body ready for Monte Carlo Rally.

Below:
The Daimler Dolphin in action. In all, around 25 Dolphins were built, but almost half of these were destroyed during the Coventry Blitz..

Right:
The Dolphin. After the Monte Carlo Rally of 1939 the car was given this new four seater body by Charlesworth of Coventry. In this form it was again rallied in the experienced hands of Bob Crouch and George Fabel.

Crouch and Cyril Simpson at the wheel), but was rebodied by Charlesworth as an open four seater tourer for the British event. It was on the RAC Rally that the team of Crouch and Fabel experienced a head gasket failure, but with the help of two works fitters, a new one was made "in record time" from Plexeal material. The Dolphin was also entered on the Scottish and Welsh Rallies.

This more sporting approach was encouraged by Edward H. Cooke - the ex-BSA man who was Daimler's Managing Director from October 1937 to February 1941. Andrew Whyte, the respected motoring historian, admired the man, saying *"He introduced up-to-date and stylish models with sports-like performance which was aptly described by the Press as 'lifting the old lady's skirts'*. This resulted in the collection of numerous competition awards in motor rallies by Daimler cars which, of course, also maintained an unparalleled record as coachwork prize-winners in such events. Official participation was undertaken by the Daimler Company, and several members of the organisation took part in rallies both as drivers and navigators. Had it not been for the coming of the war, it is likely that the Company's products would have achieved even greater success, for the competition experience gained by the Daimler team bore fruit to a greater extent with each passing season.

The EL24, Light Straight-Eight and 4.5litre Straight-Eights were continued for 1938 almost unchanged. In the August of that year, however, the Light Straight-Eight was given a larger 4litre engine, and the chassis was given the new designation E4litre. This

Below and bottom of next page:
An extract from the Twenty-Four catalogue. Several interesting models are shown, including limousines and sports saloons from such famous coachbuilding concerns as Windovers, Mulliner and Charlesworth.

Above: *George Stapley's concours winning E4 Straight Eight with pillarless sports saloon body by Vanden Plas. It was exhibited by VDP at the 1938 Earls Court Motor Show.*

for Limousine and Sports Saloon

The model shown below with coachwork by Windovers Ltd., of London, well represents the dignified yet graceful style established by these Coachbuilders. It is fully coachbuilt throughout by highly skilled craftsmen. The four doors are conveniently wide for easy access, and in spite of the low built appearance the door windows (which drop to open with easy action lifts) give a wide angle of vision from within the car. The quarter windows are also of generous size. The luxurious rear seat has side and centre folding arm-rests, and accommodation for two extra passengers is provided for by very comfortable folding chairs which fold away conveniently when not required for use. The whole of the rear compartment is sumptuously upholstered in soft cloth and a fine quality carpet is provided.

PRICE
£995

N.B.—A similar car with a folding Landaulette head is also available at £1,020.

Full accommodation for five persons is provided in the Daimler Twenty Four Sports Saloon. Front seats are independently adjustable to ensure complete seating comfort. Upholstery is of finest grade furniture hide in colours to harmonise with exterior schemes.

A pile carpet is fitted in front with a silk wool mat in rear, both with felt underlays. Safety glass throughout. Windows controlled by "easy wind" handles and adequate ventilation is always assured without draught. Concealed self-cancelling traffic indicators, box type enclosed rear number plate with reversing light automatically controlled by gear lever, stop and rear lights. Concealed dual windscreen wipers. Licence holder, interior and wing driving mirrors, cigar lighters and ash trays at both front and rear. Rear locker provides ample luggage accommodation and lid forms platform for additional luggage.

Available in a range of attractive colour schemes.

PRICE
£870

7

Right:
A beautifully restored 1939 EL24 by Windovers.

new power unit (rated at 29.8hp) offered 95bhp, and much improved acceleration. Many body styles were available including touring saloons, sports saloons and cabriolets, all varying in price from around £1,070 to just under £1,400.

On the Lanchester front, the Roadrider became the totally new Roadrider de Luxe (or LA14-2) for the 1938 season. Basically equivalent to the DB17, it had the Fifteen's old 1,809cc engine, but whereas the DB17 increased its wheelbase from 9ft 1.5ins. to 9ft 6ins., the Lanchester retained its original dimension, thus making the Daimler a slightly bigger car overall. There was also an LA14-3 launched at the same time. For export only, this model was given the option of a synchromesh gearbox, as opposed to the usual fluid-flywheel arrangement.

During late-1938 and early-1939, a number of six cylinder 24hp engines were fitted to the V-type Straight-Eight chassis to enable coachbuilders to produce large limousine or saloon bodies. Quite a few of these Type ELS24 special chassis were constructed, and Windovers in particular used them regularly. A new standard body was designed and built by Daimler themselves to fit the ELS24 chassis; the car became known as the ES24 and was famous for its sloping and rounded radiator grille - it was the first Daimler car to possess this feature.

More Royal Cars

So far, the Royal Daimlers have only been mentioned very briefly. The long-standing connection between the Daimler Motor Company and the Royal Family began in 1900, when the Prince of Wales (later King Edward VII) ordered his first Daimler car after trying one a couple of years earlier. From that date onwards, the Company was consistently honoured with Royal patronage. At an early date, King Edward VII had foreseen the enormous potential of the automobile and the important role it would play in the development of British commerce, and he encouraged the industry by setting an example in the only practical manner possible. At a time when motor cars were regarded as little more than a passing phase, His Majesty used his car on a great many occasions. He learnt to drive himself, and this alone helped to overcome much of the early prejudice shown towards the motor vehicle.

King George V did not share his father's enthusiasm for motoring as a sport, but when he came to the throne he ordered several new Daimler cars. The Royal Mews behind Buckingham Palace were converted into suitable garages, and the first fleet of Royal cars was housed there. When King George died, there were ten Daimlers in active service; five for His Majesty's use on important occasions, a private brougham, a State limousine for Her Majesty Queen Mary, a private car

Above: *Another view of the ES24, which incidentally, was used as a works rally car for the 1939 season.*

Below: *Bob Crouch drove this superb Charlesworth bodied 3.5 litre Light Straight-Eight on the 1937 Monte Carlo Rally.*

Left:
This American style 4.5 litre Straight-Eight Daimler was built for King Edward VIII, although he abdicated before actually using the car.

Above:
A procession of Daimlers used at the time of the Coronation in 1937

Below:
Another Daimler Straight Eight in competition, this time on the 1939 Scottish Rally. Despite their bulk, the big eight cylinder cars performed admirably.

for Her Majesty's use, and two cars for the use of the Royal Household. King George used his cars whenever he had to travel by road for long distances.

Almost invariably, he motored from Sandringham or Buckingham Palace to the Newmarket race meetings, and whenever he elected to stay in his suite at the Jockey Club rooms in Newmarket High Street, there was always a Daimler waiting for him outside. The Royal procession to Epsom Downs on Derby Day always consisted of Daimler cars - those constructed for King George's use had special high bodies to allow the King to wear a top hat or Guard's bearskin without difficulty.

The King's concern for the industry was well illustrated during the days immediately before the serious trade slump of 1931. The falling off in world demand was beginning to seriously affect motor manufacturing, so to help counter this, the King ordered five new Daimlers to replace those he had used previously. This gesture set a shining example by voluntarily sacrificing a very large sum of his personal civil list.

The Duke of Windsor was also a keen motorist, and almost invariably drove himself as did both the Duke of Gloucester and the Duke of Kent. When King George VI came to the throne, he already owned two Lanchester cars, one of these being a sports model which he always drove himself. It was in this car that he used to take Queen Elizabeth and the two Princesses

Above: *Another picture of CDU 202. This time on the Llandrindod Wells test on the Welsh Rally.*

Below: *George Fabel with the Daimler Dolphin following the RAC Rally.*

Above: *The 1.4 litre Lanchester Eleven ran from 1937 to 1939 and was available in two basic types; the six-light saloon or the model 442 sports saloon (Below).*

Left:
Design work on the next important Lanchester model was well advanced when, unfortunately, war broke out. This is an LD10 design dating from 1939.

Right:
The 1939 Daimler coupé (chassis no. 46123) which was once King George VI's own personal car. It is now owned by the Marquis of Bath.

from 145 Piccadilly, to his country house, Royal Lodge, Windsor Forest, nearly every weekend.

King George VI worthily carried on the tradition of his father, and as soon as he became King, immediately ordered some additions to the existing fleet of Royal Daimlers. His first new car was delivered towards the end of April 1937, and it was specially designed to be both a means of transport and a travelling office. It featured elaborate reading lamps and a small writing desk fitted into the centre arm of the back seat.

A vast number of Daimler cars were brought into service during the Coronation celebrations of 1937. It was estimated at the time, that something like 1,000 Daimlers were involved. There were 150 Straight-Eights alone, specially chartered for the use of distinguished guests. Many Daimlers were sent to the coastal ports to meet the overseas visitors and bring them by road to London. It must have been an impressive sight when, on Coronation Day, the long line of Daimler cars swept out of Buckingham Palace carrying the many Princes, Princesses, Ambassadors, Ministers, Admirals and Generals to Westminster Abbey. At this point the Daimler Company had been honoured with the patronage of three generations of Kings, even though the British motor industry was not yet 50 years old.

Above:
This 1938 Brush bodied double decker was put into service with the Coventry Corporation immediately before war was declared and was capable of seating up to 55 people. It was around this time that the first of the diesel commercials to be fitted with a Daimler engine was introduced, the first bus to have one being the CWD6. After the war this was refined into the famous CD650.

Chapter Eight

PRODUCTION FOR VICTORY

After the First World War, Britain and her allies had offered Germany much in the way of financial and other assistance, and were instrumental in fighting for world wide disarmament. Then Hitler came to power in Germany, and soon began to re-arm the country in defiance of the post-Great War Treaties. In spite of warnings from Churchill, Britain continued to disarm in the hope that her example would be followed elsewhere, but as the Treaty of Versailles was disregarded, and the Government's disarmament policy continued, it soon became clear that unless the country re-armed itself soon, its existence as a great power would be under threat. Thus, a rearmament programme was adopted and with it, what became known as the Shadow Factory Scheme.

In 1929, the Daimler Company was first commissioned by the Government to manufacture gun turrets for aircraft. At this time, there was no special urgency that the work should be completed within a certain period. The ordinary work of the Company proceeded much as usual, and nothing more than a small section of one factory was earmarked for the contract.

This, and work of a similar nature, continued for the next seven years, but in the meantime the ominous signs that Germany meant trouble were becoming increasingly apparent.

Facing Page: *A Daimler fighting vehicle in action.*

AU COIN A DROITE
HOTEL DU BRA
CHASSEUR DE
BONNE TABLE . BONNE CAVE
138

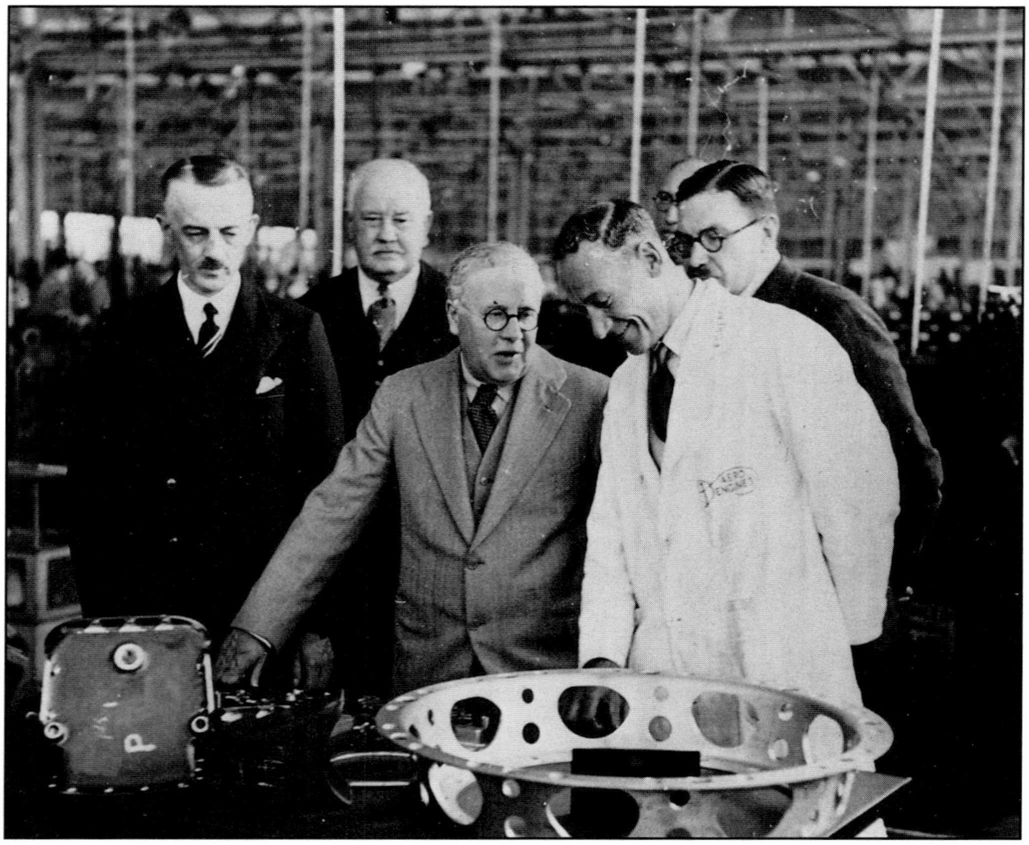

Above: *Sir Kingsley Wood, the Air Minister, on a Shadow Factory visit before the war.*

A record of Daimler's history leading up to, and production during the war years, was sent to St John Nixon by Col. George Long, the Company's long serving Advertising man, through the offices of George Hally - the Managing Director (born in 1885, and given his seat on the Board in February 1941). As these records (a 16 page document) were written up only months after the conflict by Daimler personnel, they have been quoted almost word for word, as they give the most reliable insight into the Company's war effort.

The Shadow Factory Scheme

During the May of 1936, some of the leading firms in the British motor industry were approached with a view to establishing the Shadow Factory Scheme. The idea was that new factories should be erected and plant obtained, so that if war broke out facilities for mass-production of arms and other weapons of war would be readily available. Various conferences were held, and it was finally decided that the Daimler, Standard, Rover, Austin and Bristol companies, as well as the massive Rootes organisation and the Alfred Herbert Machine Tool Co., should make a combined effort at producing military equipment in large amounts - in particular, aero-engines. There was no undue Government interference, and the companies decided between them that Daimler, Standard, Rover and the Rootes Group, should each manufacture different components for the engines, whilst Austin and Bristol would build the engines up from the parts supplied to them.

The Shadow Factories, though built by the respective companies to their own designs, were entirely Government owned. Each company was given a free hand in the control of its factory and given instructions to manage it as though it were part of its own business. All expenditure was authorised by the treasury department of the Ministry of Aircraft Production, and there was no interference whatsoever regarding rates of pay, labour employed, working hours or any other issue except for the purchase of materials, and for this authority had to be obtained first. A fee was paid to each of the companies involved for their services. These fees were small and were reduced on more than one occasion during the course of the war, but due to the complete absence of red tape and other characteristics usually associated with officialdom, the relationship between the Government and the members of the scheme was a completely harmonious one, resulting in few complaints.

The whole scheme was put into operation quickly and early in 1937, a small trickle of supplies from the companies involved began to emerge from the production lines. Vast quantities of new machinery had had to be obtained, labour recruited and the extensive organisation behind any large factory had to be arranged and set in motion.

For some time the Shadow Factory Scheme worked well, but a serious flaw came to light which necessitated a drastic reshuffle. Four of the companies involved in the project manufactured parts, while the other two firms built the engines from these components. However, if for one reason or another, one of the companies failed to deliver a particular part, there was an inevitable hold-up in supply. It was therefore arranged that Daimler and Standard should manufacture complete engines between them, and that the other companies should do the same. This way, if something went wrong, engines would be held up from only one of several sources.

It soon became necessary to provide additional storage accommodation for preserving documents, jigs, tools and manufactured parts; suitable premises were eventually found in Leominster, and at Yardley in Birmingham. It was fortunate indeed that these stores were available, as Daimler's No. 1 Factory soon got ahead of its programme, and altogether a total of 1,700 sets of surplus parts had to be sent to Leominster.

The wisdom of having extensive stores widely separated was well illustrated during the Coventry Blitz of November 1940, when over 800 sets of parts were destroyed. A further 700 sets were written off the following April, but thanks to the store at Leominster, the losses were quickly made up and the full programme was maintained. This store of surplus sets was set up without authority, and when the matter came to the ears of the Ministry, the Daimler Company was actually given a sharp reprimand for not having gained permission to buy the necessary material first!

When the Shadow Factory Scheme first started, production made slow but steady progress, and so matters continued until the Munich crisis during the summer of 1938, when an immediate conflict between England and Germany seemed, for a time, unavoidable. At once, the factories at Coventry were asked to speed up production. The immediate crisis passed, but when the war clouds once again began to loom during the

Below: *A rare picture showing aero engine production at the Daimler works.*

Above: *Originally a Daimler Shadow Factory, the Browns Lane works are now the home of Jaguar Cars Limited. This picture was taken in the mid-1970s, by which time the site had grown considerably.*

spring of 1939, it became clear that the one Daimler Shadow Factory would never suffice to meet the demands that would be made on it should a full-scale conflict break out.

During May of that fateful year, when Germany was marching through country after country, and every small power in Europe seemed on the brink of being devoured by the Nazi onslaught, Daimler was asked to build another Shadow Factory capable of manufacturing 200 Bristol Hercules aero-engine sets per month. This new factory was earmarked to produce roughly half of the Hercules engine output, but the first question to decide, was where should it be built. The Ministry stipulated that it should be within five miles of Shadow Factory No. 1 (i.e. Radford) to aid liaison, but the Daimler Company, fearing the possibility of air raids, pressed for it to be located in a totally different part of the country.

The Lancashire town of Preston was first looked at, but the Ministry refused to sanction this choice. Then came the idea of using a site in Exeter, but as soon as negotiations were opened up in that direction, the local inhabitants raised all sorts of protests, and so the original suggestion of Coventry was reconsidered. A suitable 62 acre site was found in Browns Lane, in the Allesley district of Coventry. A report was made to the Ministry and within just one week the Company had taken possession of the land, and arrangements were being made to start the building work. The total area of the site was eventually put at 844,936 sq ft, with the lay-out being the work of the Company's Managing Director, George Hally, and his son.

When the inevitable happened and war broke out in September 1939, Shadow Factory No. 2 was nearing completion. All the necessary machinery was in the process of being installed, and Factory No. 1 was increasing its output at a substantial rate. It was not long, however, before the demands of the Government forced Daimler into enlarging the accommodation space at the No. 2 site and in due course, a road which had originally run between the two main shops was covered in, and one massive unit - 1,530ft long by 400ft

Above: *This Luftwaffe map clearly shows the key bombing targets in Coventry, the Daimler works (outlined) being of prime importance.*

wide with an overhead clearance of some 20ft to allow for an overhead runway to be installed, was formed.

Just as it had been necessary to find some outside accommodation for stores and so on, when Factory No. 1 got into its stride, the same thing had to be done for the No. 2 Factory. Fortunately, as the other stores which catered for the No. 1 site still had some space to spare, only a small extra store in Bromyard was needed.

For the first year or so, Factory No. 1 was wholly engaged on Bristol Mercury and Pegasus engines, whilst factory No. 2 concentrated on Bristol Hercules units. As the needs of the armed services increased, more and more was asked of the Company, but unlike events during the First World War, there was no spectacular feature about production, everything started and proceeded in a reasonably well-organised manner. Obviously, the lessons that the Company had learnt during the First World War, proved to be of inestimable value in this latest conflict.

Soon after the war began, the Service Department was moved from its accustomed quarters adjoining the No. 2 Engine Shop, to premises at King Street and Osborne Road in Coventry. It was also found necessary for the Company to take over part of the Radford Social Club to provide extra office accommodation. The large shop which had housed the service department was then converted for the manufacture of Bren guns. For a short while, a small tennis racket factory in Nuneaton and an equally small garage were taken over for service work, but these were soon abandoned after sustaining severe damage during one of the many bombing raids.

From May 1940, the number of dispersal units which operated in connection with the Shadow Factories multiplied at a feverish rate, as war production went ahead. The air attacks on Coventry meant that all available factories, shops, garages and even private houses, stood a chance of being taken over at extremely short notice, and converted for war production of one sort or another. Early in 1941, a new dispersal unit was established in Bangor, North Wales, spread over six

Right:
The devastated main office block at Radford.

Below:
Bomb damage to the main entrance at the Radford works caused during the Coventry Blitz.

different sites. A large part of Penrhyn Castle was taken over and used as a head office, whilst premises ranging from a slate quarry to a mineral water store were put to good use. By the end of the conflict, the Daimler Co. had also commandeered an ice rink, a dye works, a club, a cafe, an hotel, several private houses, premises used in the hosiery trade, many garages and acres of park and common land for test sites.

The Blitz

During the war, Daimler shouldered an enormous responsibility. From its beginning, the Company had been engaged on motor car production in its own purpose-built factory, but almost overnight, it found itself to be the controlling power of a far flung network of factories, each one of which required administrative personnel, a huge workforce, accommodation, and everything else associated with such an immense and complex scheme. On top of this, there was always the fear of air raids. An aerial map was found in a Belgian house in mid-1939, which had previously been occupied by several high-ranking Luftwaffe Officers. It gave a birds-eye view of Coventry with the Daimler works carefully marked out. It was clear right from the start that the Daimler Co. was going to be one of the prime industrial targets.

The horrific attacks on Coventry have been well documented, and it is unnecessary to go into detail in this book except with reference to Daimler. Until the major raid on 14th November 1940, the Daimler factories were quite lucky, sustaining only fairly minor damage compared with what was to come. War-production suffered but little, and usually within a day or two, everything was back to normal. On that fateful night during November however, Coventry, the Daimler main factory and both Shadow Factories with it, first experienced the full horrors of a concentrated air attack. As the night went on, the intensity of the attack increased, until finally it must have seemed as if the whole of the City was on fire. The main factory was struck by several HE bombs, two landmines and a flood of incendiaries. Many of the departments were completely destroyed, while others were at best severely damaged. Shadow Factory No. 1 received hits from two HE bombs and literally hundreds of incendiaries, and for the first time since the air raids started, Shadow Factory No. 2 also suffered.

Even the original Daimler Works, which had been left completely in favour of the more suitable Radford Works in the summer of 1937, were destroyed during the bombing. Only the Sandy Lane Office Block remained intact, and is still in use today by Kalmar UK - the fork lift truck manufacturers.

Gradually work returned more or less to normal after the badly damaged sections had been transferred to other areas. By doing this, production was affected only slightly. As month followed month, the optimists were beginning to relax, until 8th April 1941 that is, when another bombing of even greater intensity than the last was delivered, causing an immense amount of damage and much loss of life.

In this assault, both the Daimler main factory and Shadow Factory No. 1 were singled out early in the proceedings. Another wave of attacks resulted in damage to the main works from HE and incendiary bombs. Unfortunately, the water supply gave out, and a strong wind fanned the flames; as a result, the whole office block and several of the works buildings were burnt to a shell. The No. 1 Shadow Factory also suffered badly, but the cross walls of the factory had the fortunate effect of preventing the fires from spreading to the centre of the building.

A further raid followed just a couple of days later, and many more HE and incendiary bombs were dropped on the main factory. Damage was sustained in the Bren Gun Shop and once again in the foundry, causing a number of small fires. Several bombs with delayed-action fuses were dropped, and most of these were quickly found and made safe. The following morning, a huge hole measuring some 3 inches in diameter was found in the roof and floor of the Bren Gun Shop. It seemed that a very large bomb had fallen which had either failed to detonate, or had a delayed-action fuse, which seemed unlikely. Work started straight away in an attempt to locate it, and two days later the bomb was found with its fuse lying underneath it. A shallow hole was dug beneath the bomb, but little could be done with it in this position as the bomb was found to have a new type of fuse. Blocks were quickly borrowed from Morris Motors Ltd (all the Daimler blocks of a suitable size had been destroyed), and the bomb was carefully turned over. Eventually, the 2,400lb bomb was defused, and after a total of six and a half days, was at last no longer a threat.

George Hally, the Company's Managing Director, along with some of the other Directors, calculated that approximately 418,000 sq ft of the main factory had been lost, and so the search for new accommodation began in earnest. All likely premises were thoroughly investigated, which involved inspecting around 125 factories. No less than £350,000 was spent in adapting those premises chosen by the Company,

Left:
A Works photograph of the Daimler Scout Car.

Right:
The "Light Wheeled Tank", better known as the Daimler "Dingo" Armoured Car, on test at the Radford works, being tilted at 50 degrees off the vertical. The Scout Car actually managed 60 degrees without falling over!

and the various work was distributed widely. All bus components were machined at Coventry and Wolverhampton, and the buses were then constructed at Wolverhampton. The same thing was to happen with parts for the larger Armoured Cars - for smaller models, parts were machined at Coventry, with the testing and building of the machines taking place at Uttoxeter. Gearboxes for both the large and small Armoured Cars were manufactured at Leek. Gun turrets were machined at the Wolsey Works in Coalville, the Pirelli factory in Burton and in a large house at Braunston, and were erected at the new Shobnall Works in Burton. Bren guns were also machined in Burton, while Browning guns were machined there and dispatched in parts to a factory in Leek.

In order to keep in close touch with production,

Hally moved into Densy Lodge in Derbyshire, which was more or less in the centre of the group of chosen factories. Each Factory Manager was virtually his own master, but once a week they all had to meet in Coventry to report. Remarkably, within just two months of the April raid, the production figures had risen to an even higher level than before.

Mrs Hally, the Managing Director's wife, was quick to step in and organise surgeries, canteens and welfare arrangements at all the Daimler factories and dispersal units. For three years she worked in a totally honorary capacity, but it was later discovered in the rules of the Company that nobody could be on the staff who was not in receipt of a salary. She was asked to name her price - £1 per annum, she replied - and for the following two years, she received this amount for supervising no fewer than 17 canteens and surgeries.

Daimler Fighting Vehicles

In addition to aero-engines and weaponry, Daimler continued to produce motor vehicles during the difficult war years. Two of the most remarkable and interesting vehicles of the war were the Daimler Scout, and the Daimler Armoured Car Mk.I, both of which performed admirably in all theatres of war.

During the early part of 1938, discussions had first taken place between the Army and Daimler with a view to designing and building a new type of high-speed Scout Car. The new vehicle would have to be capable of travelling over fields full of deep holes and bomb craters; it would have to be able to negotiate large ditches, fords, ponds, hedges and gradients as steep as 1:2. The fact that the vehicle would have to be capable of high speed ruled out the possibility of using tracks.

This complex problem facing the Daimler engineers, had to be solved quickly. By the beginning of June 1938 however, not only had the designs and drawings been completed, but production of the first vehicle was also well in hand. Within only six months the Scout had been designed, manufactured, provisionally tested and even handed over to the authorities for a 10,000 mile reliability trial. The vehicle survived the extensive testing and during May 1939, with war clouds looming, the first of many orders for the Scout was placed with the Daimler Company (although Alvis, who had also built a prototype Dingo Scout Car, very nearly clinched the order themselves). The first batch was delivered to the Army in December, by which time, the war had been in progress for about three months.

The Scout Car (which would become known the world over as the Daimler Dingo) was powered by a six-cylinder Daimler engine of 2.5litres. The engine had originally been destined for use in the Company's passenger cars, but when war broke out, this idea had to be shelved. The engine differed from the standard pre-war DB18 unit in several respects, having a new type of combustion head which allowed for a higher compression ratio without the fear of pinking; it also had inclined valves like the Dolphin.

A dry sump lubrication system was adopted for the Scout, as the conventional arrangement depended on a submerged circulating pump which, in rough conditions, would have been quite unsuitable. Coil ignition was provided with a governor distributor, and the whole of the ignition system was screened to avoid interference with radio transmission or reception. The carburation also had to be given special attention due to the extreme gradients which the vehicle was expected to climb. Eventually, a dust proof three stage Solex carburettor was fitted with a non-mechanical accelerator pump and a large air cleaner.

Fluid-flywheel transmission was employed, but this too was slightly modified to reduce transmission slip. This was achieved by taking away the annular guide ring of the driving member and fitting an oval oil-circulating vortex in place of the circular type. Standard Daimler practice was adhered to in the epicyclic gearbox, except that five speeds were now provided instead of the usual four. Reverse was obtained by means of a separate hand lever, and all five gears were operated by the familiar preselector method.

Coupled to the gearbox, was the differential and gear transfer unit which, in addition to distributing the drive to each of the four wheels independently, included the reverse gear mechanism. This method allowed all five speeds to operate in either a forward or backward direction. The transmission provided a constant drive to all four wheels, and the specially designed differential gear completely took away any diagonal wheel spin. The steering was arranged so that it was only necessary for the driver to move his head, and open a hinged panel to be able to drive forwards or backwards.

The frame was a plain rectangular structure with side members of uniform depth and section. In the middle of the frame, there was a forged member of I-section which was forked, allowing for attachment to the top and bottom flanges of the sides. The vehicle was laid out so that no part of the chassis mechanism would project below the lower flanges of the frame side-members, and the whole underside of the armour-

Right:
An Armoured Car chassis. Note the two steering wheels - one at each end of the vehicle.

Left:
A Daimler Armoured Car being put through its paces on one of the several test sites that the Company used during the war years.

Right:
The "Coventry" armoured car - a joint venture between two of this countries oldest car manufacturers, Daimler and Humber.

plated hull was quite flat. The approximate weight of the complete vehicle was three tons.

Until the Coventry Blitz of April 1941, the assembly of all Scout Cars was carried out at the Daimler works, but owing to the massive damage sustained during the raid, it was necessary to move production to the old Bamford & Co. factory in Uttoxeter. Production of the new Mk.I Armoured Car also had to be moved at that time, and the work, which had only just begun, was transferred to one of the dispersal units in Wolverhampton.

Although it shared a number of features with the Daimler Scout, the Mk.I Armoured Car was a far more formidable vehicle. Its design was both outstanding and original, and was one of those engineering products which, had it have been a captured foreign machine, would have caused a mild sensation. The thoroughness of its design, and its all round capability were nothing short of remarkable.

Whereas the Daimler Scout (or "Dingo" as it became known) was designed largely for reconnaissance work, the Mk.I was a much bigger vehicle and designed for use in battle. Its weight was around seven tons, and it had a wheelbase and track of 8ft 4ins. and 6ft 8ins respectively. It was a great achievement for the Daimler engineers to design a vehicle which embodied so many radical departures from accepted practice within just six months, for this was the time it took for the project to became a reality. From dawn to midnight, seven days a week, the Daimler staff had laboured to bring the vehicle into production - even the air raids were unable to stop them.

To be effective, an armoured car must solve multiple and sometimes conflicting design requirements - chiefly, it is a mobile gun, and therefore must be designed around its armaments - consisting of a fully manoeuvrable turret - and be heavily armoured. It must have exceptional all-terrain capability, provide a stable gun platform, and ideally, have the capacity to retreat in haste without having to turn on its tracks.

In the Daimler design, the engine was mounted as low as possible with a "flat belly" ground clearance, and the tail of the fluid-flywheel and the five-speed gearbox was coupled to a transfer box. From there, the drive was split into two widely separated lines which were tucked in close to the sides of the frame, and thus ran clear on each side of both the engine and the driver, allowing them to be placed low in the vehicle. Another feature of the transmission, was its central differential gear in the transfer box, which operated between the near-side and off-side wheels, and provided the vehicle with diagonal grip through fully independent suspension.

The Daimler could perform a hasty retreat, for besides the main steering wheel and controls for the driver, there was a second set of controls (except for a gear lever that is) at the back of the turret, which allowed the Commander to take charge and drive the machine backwards. The transmission provided five speeds in either direction, and the maximum speed was in the region of 50mph.

It is easy to understand that a vehicle weighing around seven tons must have a very low bottom gear ratio for emergency purposes, and this was fixed at 80:1. The torque which can be generated by an engine of 100bhp through an 80:1 gear reduction, is of course, very high indeed, and to put it through large reduction gears and axle shafts, would have meant using components of a large size and weight in order to obtain the requisite strength. So an arrangement, quite new to the industry, was employed - an epicyclic reduction gear was contained in each hub, and this gave a final reduction of something like 31:1. By this method, it became possible to avoid overloading the axle shafts and Tracta joints.

At each corner of the vehicle, as it were, there was a fixed spiral bevel reduction gear, the pinion of which was driven by its respective propshaft, while the crown wheel was coupled to a Tracta cv joint contained in a spherical casing. A glance at the front-wheel drive, shows how a short enclosed shaft ran outboard to a second enclosed Tracta joint. This was attached to the sun gearwheel of the epicyclic reduction gear in the hub of the road wheel. These reduction gears were detachable without removing the wheel.

The engine was a six-cylinder unit of 4,095cc which developed around 100bhp at 3,600rpm. Twin Solex carburettors were used, and these drew their supply of air through an oil bath-type air-cleaner, delivering mixture to the inlet valves through a manifold heated and temperature-controlled by water from the radiator by-pass hose. Bolted to the casing of the fluid flywheel, was a Wilson type gearbox of the epicyclic self-changing variety which provided the vehicle with its five different speeds. Braking was via Girling discs.

In construction, the frame of the Mk.I resembled a flat-bottomed punt made of armour plate. All the mechanism was thoroughly protected inside the machine, and the bottom could withstand the very roughest of treatment. Towards the middle, there was a blister on each side, each of which contained a fuel tank of 36 gallons. In common with the frame, the hull

Right and below:
Two photographs showing Daimler fighting vehicles in action.

was made of welded armour plate, and the two were bolted together to form a unitary construction of great rigidity. When finished, the overall length of the vehicle was 13ft, while the height was 7ft 4ins, and the width was 8ft.

In 1942, the Army asked the Daimler Company to design a scheme for sealing the Scout and the Armoured Car, so that they could wade ashore onto beaches under their own power after being dropped off by landing craft. The original system was so successful, that more rigorous conditions were imposed. Eventually, the Armoured Car was developed so it could make its way through about 5ft of water (with an 18inch wave allowance in addition) while the Scout could quite easily negotiate 4ft of water. Subsequently, Daimler was asked to design and fit prototype equipment for Staghound armoured cars and Sherman armoured recovery vehicles, as well as being asked to supply information on sealing to the other firms building armoured machines.

Writing about these Daimlers in the *Imperial Review* of January 1944, Ralph Fielden noted: *"Summed up, my impressions of the two Daimler vehicles, over the most severe course I have covered, is that Britain has nothing to fear from the competition of any country so long as her manufacturers produce such outstanding vehicles."*

Just before the end of the war, Daimler and the Rootes Group decided to design and build a new armoured car between them, to replace the Daimler and Humber machines. The work was co-ordinated by Humber, although both parties produced prototypes, the first of which were on test in 1944. Christened the "Coventry" Armoured Car, the machine was powered by an American Hercules six-cylinder petrol engine giving 175bhp, and featured a more conventional four-wheel drive system than the old Daimler. Large contracts for these new fighting vehicles were issued, but by the time that full scale production could get under way, the war had ended, and the contracts were cancelled with only a few ever being made. British Forces never used the Coventry in battle, although the French Army used them in their Vietnam campaign.

In any case, several thousands of the Daimler Dingo Scout Cars and Mk.I Armoured Cars were turned out during the war (6,665 Scouts and 2,674 Armoured Cars), but it must not be assumed that this was all that the Company did....

Other War Work

The manufacture of aeroplane gun turrets, as mentioned earlier, began in 1929. Later on, the Company built turrets for the Blenheim aircraft as the result of a contract awarded by the Bristol Aeroplane Co. As soon as the Shadow Factory Scheme got under way, the Ministry of Aircraft Production placed several large orders, and during the course of the war, huge numbers of turrets were turned out for various flying machines, the total quantity running well into five figures.

During 1940, the production of gun turrets moved from Coventry to Braunston House, in Braunston, where some 125 people were employed. In peace-time, the building was used as a cafe but, for a time, it found itself transformed into a very useful 'factory'. A short while later, the operation was moved once again, this time to the Wolsey Underwear factory at Coalville, and then again to another works belonging to Wolsey at Ratby. The increase in turret production was due largely to the taking over of the Pirelli works at Burton-on-Trent in 1942, and for the building of a new factory in Burton known as the Shobnall Works. Orders for ancillary equipment from the Ministry of Aircraft Production were taken and processed on a vast scale - hydraulic boosters, mountings etc., were turned out in great numbers. These orders were followed in 1941 by a request to overhaul and repair the Daimler-built turrets, and literally thousands were serviced by the Company.

Twentythree parts of the famous Bren gun were manufactured by Daimler. The main body of the gun, as it arrived at the factory, weighed 41lbs, and was nothing more than a rough forging. After 220 machining operations, it was despatched weighing only 5lb. This work was originally carried out at Coventry, but after the 1941 Blitz it was transferred to the Eatcough's Boot & Shoe Factory at Burton. By the end of the war, production numbers for Bren gun parts had run into six figures.

Certain parts for the Browning gun were also machined by Daimler. Again, initially this took place at the Coventry works, but two years after the start of the war, operations were moved to an old hosiery factory in Leek. In 1940, a department had been set up at the main factory to recondition Browning guns, but during the air raids of November 1940, the department was totally destroyed. A similar fate befell a programme for manufacturing components for Boulton-Paul airframes, such as wings etc., at Coventry. Work had just started after many weeks of preparation, when the same air raid in November reduced the contents of the workshop to rubble.

The epicyclic gear, as found on the Daimler-Lanchester cars of the time, was to play an important role in the control of tanks during the war. As there are no road wheels on tanks, the only way to steer them is by speeding up or slowing down one of the tracks, and this was made feasible by a Wilson-type epicyclic gear. It was located between the main transmission unit and each track, and proved to be very effective in operation. After being built and tested at the Daimler main factory, the units were then despatched to the tank manufacturers for installation in Crusader, Covenater and Cavalier tanks. More than 2,500 transmission units were delivered during the six years of the war.

Daimler also took an active part in the production of both rocket projectors, and special fuses for these which were of a very unusual construction, demanding a high degree of accuracy in their manufacture. The work on the rocket projectors, consisted mainly of designing and manufacturing prototypes, and was first undertaken in 1938. Many prototypes were built and tested until one was finally standardised, and handed over to be produced by the Birtley Company - a Daimler subsidiary, absorbed in March 1939. The design and construction of the shell fuses was eventually put out to another firm after the special workshop at the Daimler works was bombed,

Right:
A Daimler propaganda poster showing armoured cars in the Victory Parade at Tunis, May 1943.

Below:
A Daimler Scout and Armoured Cars in a convoy during a victory parade.

moved and bombed again in the Blitz. The Birtley Co. turned out vast quantities of items during the war, both for the Admiralty and for the Ministry of Supply.

It is not commonly known that Daimler from as far back as 1929, had been the chief sub-contractor for the Bristol Aeroplane Company, producing certain engine parts. This connection was to continue throughout the war, but on a far larger scale than ever before. The result of this private contract, was that a staggering five million components were delivered up to the spring of 1945. Again, before the war, Rolls-Royce aero-engine parts had been manufactured under sub-contract, and orders for these increased as the war raged on. Up to forty items per week were delivered to Rolls-Royce, all of them being fairly large parts, such as propeller-shafts.

Daimler's involvement with the aero-industry can be gauged from the following production figures;

Bristol Hercules 14 cyl aero-engines	*20,302*
Sets of spares for the above	*3,717*
Bristol Mercury 9 cyl aero-engines	*18,306*
Sets of spares for the above	*4,237*
Bristol Pegasus 9 cyl aero-engines	*12,195*
Sets of spares for the above	*1,495*
"Power Egg" assemblies	*7,183*
Sets of spares for the above	*263*
Miscellaneous aero parts	*10,971,000*

The Company also landed a machining contract for drop-bar bomb release gear. This was used on virtually every type of bomber, and during the course of the contract about 300,000 sets were produced. A great deal of work was carried out for the Lockheed Hydraulic Brake Company too - undercarriage jack assemblies, flap jack and valve assemblies, and brake accumulators were just a few of the aircraft components that Daimler supplied to them.

During the early days of the war, the Company's Service Department was converted to undertake the repair and overhaul of vehicles belonging to the Armed Services. The depots in Leeds, Edinburgh, London and Newcastle, were all equipped to handle the repair of such machines as RAF tenders, fire tenders, mobile cranes, water and petrol trailers, the Daimler Armoured Cars and Scouts, and anything else that happened to come along. By the end of the hostilities, well over 2,000 vehicles had passed through the Daimler service stations.

Daimler were also involved in the *"Publican"* project, but as this was *"very much on the secret list"* at the time, and the author has been unable to find out more, little is known about it. Indeed, it is not even known which field of activity it involves.

More familiar Production

By 1942, a serious shortage of public passenger transport vehicles had arisen. This was due largely to enemy action, although the fact that war production came first meant that the replacement of worn out buses was on a very restricted basis. The shortage gradually began to have an effect on the war effort, because of the difficulty of moving employees to and from their work. In March 1942, under the authority of the Ministry of Supply, the Company once more started to produce double-decker bus chassis. The development of the Daimler 8.6litre heavy oil engine, which had been interrupted by the war, was recommenced, and by the early part of 1945, the results proved good enough for the Ministry to double their original order, and to specify the six cylinder Daimler power-unit for the increased number involved. No fewer than 800 chassis were built during the war period.

A number of Royal cars were also supplied, including four cars for King George VI and a 2.5litre DB18 for HRH Princess Elizabeth. All the King's cars were built on the long wheelbase 4.5litre Straight-Eight chassis, although the first two (ordered in March and October 1939 respectively) were badged as Lanchesters. The first, a landaulette, was delivered as war broke out, and the second, a limousine, arrived at the Mews early in 1940. The coachwork was by Hoopers in both cases, as it was on the two armour-plated limousines which were delivered during September 1941. These two particular limousines should have been delivered much earlier, but unfortunately, the original chassis were completely destroyed in the air raid of April 1941. The DB18 was in fact an 18th birthday present for Princess Elizabeth from her father, who bought it for her in 1944. This new saloon was registered JGY 280, and the Princess retained this plate for a number of her future Daimler cars.

During the last year or so of the war in Europe, the Daimler Company was told by the authorities that there was an urgent demand for some of the bigger Daimler cars to be used by ambassadors and officers of the Navy, Army, and Air Force (there was also an armoured car similar to the Royal vehicles built for the Commander-in-Chief of the American Army). This request was made when the vast majority of Daimler cars had been converted for war purposes, or were awaiting disposal after having sustained damage in the

Above: *The troops move through one of the battle scarred cities of Continental Europe. Daimler Armoured Cars are bringing up the rear.*

Below: *The end of the war came as a great relief to everyone - these people are certainly happy enough knowing that it is all over.*

Coventry Blitz. No ordinary touring cars had been built since the war began, and so it was difficult to respond quickly. However, the necessary number of cars was eventually found, and all were duly handed over to the authorities, either for immediate use, or for service in what was to become occupied Germany.

The number of factories, depots and stores operated by the Daimler Company on war production amounted to 55. The total number of employees over the war period was approximately 17,000, and their effort resulted in an output valued at no less than £62,242,768. It was an incredible effort, but one that would sustain the Company for a number of years.

Hooper & Co. (Coachbuilders) Ltd

Hooper & Co., who date back to 1807 and gained their first Royal Warrant in 1830, fell upon hard times, and were taken over by Daimler in 1940. The Company not only inherited a fine business with a list of patrons "unequalled by any other coachbuilder in the world", but factories at Acton and Chelsea, and another prestigious coachbuilder - Barker - whom Hooper had absorbed in 1938.

Barkers had been established for nearly a century longer than Hooper (they were founded in 1710), but largely relied on reputation to bring in new business. Sadly, this was a classic case of a company resting on its laurels for far too long. Having been associated with firms like Daimler and Rolls-Royce since their respective births, the high point of their career was perhaps building the body for Captain Malcolm Campbell's "Bluebird" - the Napier-engined machine which took the LSR record to over 200 mph in 1928.

Although Hoopers carried on much the same - business as usual in effect - the Barker name would soon begin to find itself being applied to standard Daimler coachbuilt cars and, in fact, all post-1945 Barker bodies were built at the Radford Works. For many years, however, these two fine concerns continued to be closely connected with the Company.

Chapter Nine

THE DOCKER YEARS

Sir Bernard Docker had joined the Daimler Board in July 1940. His father, Dudley Docker (b. 1862), had been on the BSA Board since 1906, and was made a Director of the Daimler Company in 1910. Although he resigned from this position two years later, he continued to take a great interest in the business, rejoining the BSA Board, and retaining his seat until his death on 8th July 1944. Once described as *"one of the best commercial heads, if not the best, in the country"*, he is rarely given the credit he deserves for his unique contribution to British industry.

Before he died, Dudley Docker saw to it that the association between the Docker family and BSA/Daimler continued by dismissing Sir Geoffrey Burton (Daimler's Chairman since 1936, and Managing Director of the BSA Group). After spending a weekend with Docker, he was told *"he simply had to go"* in order that his (Docker's) son could take his place as the MD of the entire BSA Group.

Sir Bernard Docker was born in 1896. He was educated at Harrow, and on leaving joined the Metropolitan Carriage Wagon & Finance Company of Birmingham (a BSA subsidiary once owned by his family), put there by his father in order for him to avoid conscription during the First World War by working with munitions. In 1920, Sir Bernard made a business trip to South America and shortly after his return, became the Deputy Chairman of Metropolitan.

Facing Page: *Daimler Conquest, 1954 Monte Carlo Rally. Seven Conquests were entered in 1954, of which five finished in the top 100.*

73
RALLYE MONTE-CARLO
OHP 831
73

Eventually, thanks to his father's intervention, Sir Bernard would become Chairman of a number of important industrial concerns, one of which was the BSA Group. Elected to the Board in 1939, he was Chairman in the following year, and Managing Director in 1944. Docker continued as Chairman of the Daimler Company for well over a decade, but this was to be a period which was not without its share of problems.

With the war well and truly over, and with rumours of the Company ending car production quelled, Daimler could at last get back to the business of building cars. First though, like so many companies in the industry, it would have to rebuild and re-equip its factories. With this work in progress, Daimler announced that it would shortly be introducing a three model range; an improved version of the pre-war 2.5litre Fifteen (DB18), the six-cylinder 4.1litre DE27, and the flagship - the DE36, which closely followed the concept of the superb pre-war Straight-Eight.

The first couple of years after the war were not easy for British car manufacturers. They had to make the best of old machinery and try to return to their normal peacetime production as best they could under the many restrictions enforced by the Government.

Virtually every department at Daimler had been disrupted by the war. Some had been only partially destroyed, but others had been put out of action completely. The Company also had the added problem of having so many Shadow Factories under its control. Its staff and equipment had been spread far and wide during the hostilities, which made immediate post-war production all the more difficult.

However, it could just as easily be said that the war was good for Daimler. In the immediate pre-war years, although income from sales was growing steadily following the Depression, 1935 saw a loss, despite selling £170,000 worth of vehicles. After a peak in profits of just £10,000 following the launch of the Straight-Eights, both sales and profits dropped dramatically. Had this situation continued, Daimler may well have been out of business by the early 1940s. Looked at in this way, the massive income from war work almost certainly saved the Company.

Eventually, as things began to return to normality, the Company moved its operations back to Coventry. The Radford works were rebuilt and the new factory at Allesley was put to good use. The fiftieth anniversary of the Daimler Company was celebrated on 14th January 1946, and a Jubilee Luncheon was held under the presidency of the Chairman Sir Bernard Docker at the Savoy Hotel in London. Over 200 people were invited, and in proposing a toast, Lord Brabazon of Tara pointed out that it had been the loyalty and enthusiasm of those in charge of the Company, which had built it up from its small beginnings into the huge organisation it was at the time, becoming something of a national institution.

Above: *Sir Bernard Docker.*

The Post-War Range

The post-war DB18 went on sale during February 1946 and had a much improved engine. Experience gained with the 2.5litre engine which powered the Daimler Dolphin, and with the military Scout Car power-unit, resulted in several modifications to the cast-iron cylinder head. This new head had pushrod operated valves set at an angle of 15 degrees, and this, in conjunction with reshaped combustion chambers and ports, allowed the compression ratio to be increased to 7:1. A new induction system was also introduced which allowed the engine to warm up more quickly. All these improvements resulted in the power output being raised from 64bhp to 70bhp, with fuel consumption being put at 22mpg for the new model as against an 18mpg average for the pre-war version. A slight modification on the fluid-flywheel, again brought

Left:
The Board pictured at "Mackadown" in June 1947. Sir Bernard is in the centre of the bottom row, and another important figure is that of George Hally, seen on the far left at the back.

Right:
The Daimler DB18 2.5 litre six light saloon by Mulliners of Birmingham, announced only days after the war.

about through experience during the war, allowed for better acceleration at low speeds, but otherwise, the specification for the rest of the basic chassis had changed very little from that of its predecessor.

It was left to Mulliners of Birmingham to build the standard six-light saloon body, and this was the last Daimler production car to be equipped with a sunroof until the top-of-the-range models of the early 1980s. Running boards were now a thing of the past on the DB18, and by using steel body members and laminated timber, it became possible to reduce the thickness of the body sides, thus giving a fairly substantial gain in passenger accommodation. Thinner pillars were also employed, resulting in better visibility without sacrificing the cars immense strength. The interior was, of course, trimmed in the traditional leather and polished wood.

Quite a few perfectly standard post-war DB18s carried a coachbuilders name plate, although the specialist was frequently only responsible for the painting, trimming or general finishing of the vehicle. There was, however, an attractive version of the model offered in four seater two door drophead coupé form. The standard 70bhp chassis and power unit was employed, and there were two common variants - one was the Barker-bodied version, while the other was by Tickford of Newport Pagnell. The hoods on both cars were spring assisted, and folded into the recess provided behind the rear seat; it was also possible to obtain an intermediate de ville position. In 1946, the price of the Barker drophead coupe was put at £1,425 9s 5d including purchase tax, and the Tickford car cost £85 more in the basic price.

The DB18 was of course a pre-war carryover, but the DE27 was most certainly not. This was a completely new car and because of this, did not find its way onto the market until the latter part of 1946. The DE27 was to take the place previously occupied by the Twenty-Four before the war, designed to bridge the gap between the Company's smaller, more moderately powered models and the large limousines and State carriages for which the marque had become world famous. The DE27 was the Company's first post-war mid-range model.

Above: *A Barker bodied DB18 convertible, as built between 1946 and 1950.*

Above: *A rare drophead coupé on the DB18 chassis by E.D. Abbot of Farnham, this example dating from 1947.*

Below: *There was also a Tickford version which was equally popular at the time.*

Below: *A DE27 with limousine coachwork by Barker.*

It was decided to utilise the 4,095cc engine which had powered the Daimler Armoured Car so successfully throughout the war, and to install it in a new chassis capable of carrying a coachbuilt body which would be equally suitable for both social purposes in town, or for long-distance travel across country. The power-unit needed some modification before it could be used for private cars - a tank engine in a Daimler car would not have gone down too well, as one can imagine!

The cylinder head received the same improvements as those made to the post-war DB18, and this allowed a 6.3:1 compression ratio. The manifold and induction system also followed the trend set by the 2.5litre car, but on the DE27, two downdraught SU carburettors were used instead of one. Eventually, the engine developed 110bhp.

Both the engine and gearbox were mounted and supported on rubber blocks, and to ensure that no engine vibration could be transmitted to the chassis, there were no centre bearings on the propshaft. Instead, to keep the shaft in proportion, there was a 15 inch extension fitted to the rear axle.

To call the Twenty-Seven a mid-range model is in fact quite ridiculous, for at the time of its announcement, the DE27 was one of the largest production cars in the world. During 1938 and 1939, the average wheelbase for the British cars tested by *Motor* magazine (exceeding 25hp) was 10ft 6ins. - the DE27 had a wheelbase of 11ft 6ins. This had the effect of giving the car an enormous amount of passenger accommodation, with some 70% of the area within the wheelbase being available for this purpose. However, a careful overall design by Cyril Simpson made sure that the vehicles proportions were never out of place, and the new rounded radiator shell (also fitted to the DE36), gave the car a very dignified frontal appearance as well.

The DE27 chassis was exceptionally robust, allowing the independent front suspension to work properly. In addition, because the chassis was so deep,

Right:
The well known Lanchester enthusiast, the Jam Sahib of Nawanger, had two DE27s built into Lanchester all weathers by Vanden Plas in 1946. Both were finished in pale blue.

Above: *A Windovers limousine, design no. 116. The first DE27 to have this body fitted was chassis no. 51042, which was delivered to Windovers in February 1946. Towards the end of the year, the first Straight-Eight was bodied, built to the same design.*

Above: *The Windovers four door saloon, design no. 136, was suitable for either the DE27 or the DE36 chassis. The first DE27 to carry such a body was produced mid-way through 1947, whilst the first Straight-Eight was built towards the end of the year.*

even with the rear wheels at the limit of their deflection, the propshaft always remained below floor level, aided by inclining the rear axle and engine slightly downwards. Rear suspension was via semi-elliptic springs, Luvax-Girling shock-absorbers and an interconnecting torsion anti-roll bar.

Steering was by Marles roller gear, but instead of employing the normal method of kingpin inclination to provide castor angle, the geometry was changed slightly to make the car significantly more controllable than its predecessors. Another new feature was the braking system; for the first time on a British production car, a Girling hydromechanical arrangement with servo-assistance was provided. With this system, the front brakes were actuated hydraulically, whilst rods and levers operated the rear brakes.

In continuation of the Company's pre-war tradition, the DE27 was also sold in chassis only form to enable customers to have the coachwork of their choice fitted. When the model was first announced, this option was made available at a list price of £1,200, although by the time that production had got underway, this had risen to £1,400.

The post-war Straight-Eight (or DE36) sported a virtually identical chassis to the DE27, except that it had a slightly longer 12ft 3ins. wheelbase, and was fitted with a magnificent eight cylinder engine of 5,460cc. However, the cylinder dimensions of 85mm x 120mm were the same for both the six and eight cylinder power units. With the extra two cylinders, maximum power was raised to 150bhp at 3,600rpm, and maximum speed was in the region of 80mph.

The new car clearly impressed most motoring journalists with its sweet running and handling. In fact, their comments were even more complimentary than those which had been bestowed upon the Twenty-Seven, despite the fact that it represented *"a small fortune in money."* In addition, it soon became obvious to the coachbuilders of the time, that the DE36 chassis offered almost unlimited scope. Virtually all of the standard bodies produced by Hooper, Windovers and Freestone & Webb, were designed in such a way that they were suitable for either the DE27 or the DE36, but it was the latter which tended to attract the coachbuilder's eye. Accordingly, during the seven years of the models production, some of the most elegant bodies ever built were mounted on this new Straight-Eight chassis.

The DE27 also attracted some fine coachbuilding; for example, the famous Vanden Plas concern built a pair of identical all-weather Lanchesters on the DE27 chassis. These cars were ordered by the Royal House of Nawanger for the use of the cricketer, "Ranji". The bodies (4031 and 4032, mounted on chassis 51049 and 51050 respectively) cost £1,695 each, and these were the only post-war Daimler or Lanchester bodies to be built Vanden Plas until the advent of the DS420 limousine in the late-1960s. This distinguished customer actually owned another pair of DE27s, although both of these were fitted with standard Hooper six-light bodies.

The standard Hooper four-light and six-light limousines usually featured a leather trimmed bench seat fitted in the front compartment, capable of accommodating the driver and two passengers. In the back, there was seating for three passengers, as well as occasional seats for a further two. The majority of these cars had electrically operated divisions, and separate electric motors were employed for the rear window blind and the driver's door window - some cars were also equipped with division blinds for extra privacy. Separate front and rear heaters could be built into the car to order, and a radio and speaker could be installed into the back of the division between the occasional seats. With the six light Hooper body, there was no pillar between the rear door window and the quarter light, making visibility quite exceptional.

Hooper built 114 bodies on DE36 chassis, many of which were significant. Some of the earliest Hooper bodied DE36s went to customers such as HM the Queen of Holland, the Prince of Monaco, HIH Emperor of Abyssinia, the President of Czechoslovakia, and HM the King of Afghanistan. A number of cars followed for use by our own Royal family and HM the King of Thailand, not to mention the lavish Docker Daimlers which were the stars of Earls Court for a number of consecutive years.

Above:
When the post-war Straight-Eight was first put into production, this six-light limousine body was considered to be the standard offering from the famous Hooper coachbuilding concern.

Right:
Brian Smith's superb Freestone & Webb close-coupled limousine. This was the last such body to be fitted to a DE36 chassis, with the car registered in 1950.

Most post-war Straight-Eight Daimlers had the spare wheels mounted on the wings under specially designed covers, and the bonnet sides featured chrome-plated slats. These small details generally distinguished the DE36 from the DE27, but unfortunately, the practice concerning the side grilles was not always consistent either way, and so cannot be relied upon as a means of positive identification.

Windovers were another famous coachbuilding concern asked by Daimler to design and build bodies suitable for both chassis. The firm's beginnings can be traced back to Huntingdon in 1796. Windovers, by now based in London, had constructed high quality bodies for a considerable number of Daimlers before the war, so it was only natural that the firm were given further orders. In 1946, Windovers were asked to produce 50 limousine bodies (to their design no. 116), which would be suitable for either of the two larger Daimler models.

In actual fact, only 30 and a prototype were ever finished. Five of these were fitted to the DE36 Straight-Eight chassis and the remainder mounted on the DE27 chassis. The very best seasoned ash was used for the framing, strengthened where necessary by steel body plates, and hand formed sheet aluminium was employed for the whole of the exterior panelling - in common with Hooper, and Freestone & Webb, the Windovers cars featured concealed running boards.

Inside the limousine, the seating arrangements were much the same as those seen in the Hooper examples. A spring blind operated by the driver was fitted to the rear window and there was a sliding glass division. All interior woodwork was in walnut, and Windovers received a payment of £688 for each complete body.

The first of the Windovers limousines was started in February 1946, and the last one left the works during the early part of 1949, both mounted on the Twenty-Seven chassis. The five Straight-Eight Windover limousines were constructed between the end of 1946 and the autumn of 1948.

Windovers offered the only owner-driver saloon on the DE27 chassis in standardised form. Daimler ordered a total of 25 bodies to this design (no. 136) and once again, it was suitable for either of the two larger chassis types. In most respects, the saloon had much

Above: *When the LD10 was eventually built, the first were manufactured during April 1946. Note the Briggs pressed steel body.*

Below: *The Royal party about to board one of the DE36 Hooper landaulettes during the 1947 Tour of South Africa. Two cars were delivered to the South African Government late in 1946 for their Majesty's use - one on chassis no. 51156, the other on chassis no. 51158.*

the same specification as the limousine, except of course, the removal of the division. However, a division could be fitted if the customer wanted one, and in all cases, a sunroof was supplied as standard - the upholstery and woodwork was left very much to the customer's choice. The saloon was originally built as a six seater, but where a division had been specially ordered, two occasional seats were generally included, making the car into an eight seater. Alternative seating arrangements were also made available, such as a separate seat for the driver.

Of the 25 bodies ordered by Daimler, a total of 20 were completed. Of these, all but five were mounted on the DE27 chassis. The first Windover saloon built on the Twenty-Seven to design no. 136 was started in May 1947, while the last one was finished in May 1949. The five DE36 saloons by Windovers were built between September 1947 and December 1948. Windovers received the sum of £753 for each completed body, plus an extra £70 if a division had been fitted.

Freestone & Webb also designed bodies for both

Left:
A four-light limousine by Hooper on the Straight-Eight chassis, circa 1947. Very few four-light versions were ever constructed.

Left:
Another Hooper limousine, this time a six-light. This contemporary picture was taken in North America.

Above: *A Barker Special Sports - around 500 were produced between 1948 and 1953. They looked equally beautiful, either with the hood down...*

....or with the hood up to beat the British weather. Many wre given two-tone colour schemes (Left).

Right: *One of the rare fixed head coupé versions of the Special Sports, the work would have been an after sales commission, Appleyards of Leeds being one of the better known convertors.*

the DE27 and the Straight-Eight. Their designs had always been attractive, and these latest styling masterpieces were no exception. The razor-edge bodies were typical of this coachbuilder's work. In total, around 20 of the larger post-war cars carried Freestone & Webb bodies, at least ten of which were limousines mounted on the DE27 chassis. There were also one or two owner-driver vehicles built without partitions, and two very special close coupled sports saloons on the DE36, the details of which we shall go into later.

One or two DB18s and several DE27s carried coachwork by Charlesworth. It is quite possible that a few Straight-Eights also received bodies from this Coventry based concern, but as all of Charlesworth's records were destroyed when the company closed down its works, it is unfortunately impossible to give any definitive answers.

A Post-War Lanchester

The first Lanchester to appear after the war was the Ten (or LD10). The LD10 had been in development since before 1939, but obviously the war had interrupted this work. In all, 25 prototype LD10s were produced before the war, including two coupé versions and a van, so that the model could be launched for the 1940 Model Year. When announced in December 1945 (it finally went into production in April 1946), the differences between the pre-war prototypes and the post-war cars were easily identifiable.

The earlier machines were fitted with the flat radiator used at the time, and they also featured quite shallow headlamps built into the front wings. On production cars, the radiator was a pointed affair, similar to the contemporary small Austins and Fords, and the headlights were mounted separately on top of the wings. The LD10 was a refined little car with a four cylinder ohv engine of 1,287cc capable of producing 40bhp at 4,200rpm. Fluid flywheel transmission was a standard feature, as was an independent coil spring front suspension.

At this time, the only body available was a six-light saloon built in pressed-steel by Briggs of Dagenham.

Left:
The original and massively impressive Green Goddess which made its debut at the 1948 Earls Court Motor Show.

Right:
Another view of the Green Goddess showing the interior; the facia was particularly stylish.

Right:
Green Goddess with the hood up. Using the Daimler staff behind the car as a guide, the immense size of the car can clearly be seen.

Below:
Early in 1949, Freestone & Webb completed this fine body on the DE36 Straight-Eight chassis (no. 51238) for a customer in Jamaica.

Before the war, Briggs had been asked to put a drophead version forward for approval, although all the prototypes had all been bodied by Barkers. The chassis was sold separately to coachbuilders, but it was rarely used, its cost being prohibitive on such a small car. As a Briggs saloon, its price on introduction was set at £595, plus a purchase tax of £166, well above its immediate competitors such as the Morris Series M and Wolseley Ten.

In addition, the original LD10 may not have been the prettiest car ever to grace the Daimler-Lanchester showrooms, but it did at least keep the Press happy with its spirited performance. In 1946, William Boddy, the long-serving Editor of *Motor Sport* magazine, said of the car: *"When the new Lanchester Ten came along for test, we anticipated refined motoring in comfort and safety. We had not been in the car very long before we were enthusiastically expressing our surprise at the performance of this modestly dimensioned car and the way it could be handled."* He added that, *"its steering and suspension combine to provide handling qualities which those who are habitually in a hurry can really enjoy."*

The Royal Tour

1947 was very much a Royal year as far as the Daimler Company was concerned. In the latter part of 1946, preparations were made in earnest to have several cars ready for the forthcoming Royal Tour of South Africa. As well as many standard DB18s, several coachbuilt vehicles were also supplied for the tour.

The South African Government ordered five DE36 Straight-Eight Daimlers with bodies by Hooper. These included two landaulettes (chassis 51156 and 51158), two limousines (51159 and 51160), and an all-weather car built on chassis no. 51161. Another limousine, with coachwork by Windovers, was ordered by the Transvaal Administration for use on the tour, and this carried chassis no. 51180. In all five of the Hooper-bodied cars, both the front and rear seats could accommodate up to three people, and behind the electrically controlled divisions in the limousines and landaulettes, a further two passengers could be seated on the forward facing occasional seats. All windows were electrically operated, and the interior appointments were naturally finished to the highest quality. The upholstery was in a plain grey ribbed cloth with a matching grey headlining, and silk blinds were fitted to the door windows and division - the landaulettes featured a special transparent roof panel which could be covered by another roller blind when not in use. The coachwork was dark blue over black, with fine red coachlines applied along the waist of the vehicle.

All of the Hooper cars were finished and ready for use by the end of November 1946. For a time, they were proudly displayed at the works before finally being shipped to the Royal Albert Docks along with the standard 2.5litre DB18 saloons for their long journey south on the "City of Madras". The Windovers car was built to their standard design (no. 116), and carried the appropriate registration; "SA 1". The car was beautifully finished in black with a red leather interior, and could seat up to nine people.

The Royal party sailed from Portsmouth on 1st February 1947, in Britain's newest battleship, HMS Vanguard. The King, Queen and two Princesses were to visit the Dominion, and the King was to formally open Parliament on 21st February 1947. The Chairman of Daimler, Sir Bernard Docker, had been requested to oversee the management of the cars during the tour, with the responsibility for the day-to-day running of the Daimlers being left in the capable hands of Bob Crouch. So, early in January, Docker drove down to the South of France to join his yacht "Shemara" for the voyage to Africa. However, during the voyage, Sir Bernard's mother died and he had to return home, leaving Bob Crouch to make all the arrangements for the cars.

Crouch, whose father had built the Crouch car in Coventry between 1913 to 1927 (and three wheelers before that), had joined the Company before the Second World War. He had made his name rallying the big Straight-Eights of pre-war times, and was gradually given positions of more and more importance. He stayed with the Company even after the Jaguar take-over as Sales Manager of the Daimler bus division.

However, the Tour, which ended on 27th April, was a great success for all involved. The Royal Tour Fleet at times comprised of 31 vehicles, which between them covered a total of nearly 200,000 virtually trouble-free miles. At the end of the Tour, Bob Crouch received a letter of thanks and a pair of gold cuff-links from the King and Queen, who were *"deeply grateful"* for his sterling work.

Above:
A Hooper landualette used on the 1954 Australian Tour, photographed recently at the Jaguar works at Browns Lane. It will be seen that the tour should have taken place in 1949, and so all of the larger cars were built with this date in mind.

Left:
This Freestone & Webb close coupled sports saloon was the sister car to the one that went to Jamaica, and was exhibited at the 1949 Motor Show.

Above: *In 1949, Hooper limousines were given a much neater and more attractive rear end. This particular example was actually exhibited at the 1949 New York Show, held in the February of that year.*

Below: *This Straight-Eight landualette was built for the King of Siam by Hooper in 1949.*

Above: *For such large cars, the Daimler Hire limousines were well proportioned - around fifty were built.*

New Royal Daimlers

There were also some new Royal Daimlers for use at home in 1947. There were two Straight-Eight landaulettes for King George VI, and a limousine for Queen Mary built on the DE27 chassis. In all three cases, the coachwork was by Hoopers, who managed to have the cars ready just in time for the King and Queen's return from South Africa.

The King's cars were delivered at the beginning of May, and were much the same as the landaulettes used on the Royal Tour. Chassis no. 51172 and 51177 were used, and these carried the bodies respectively numbered 9336 and 9335. Their appearance was very similar to the Tour cars, except for the forward hinged front doors and the famous Royal colours that were used for the exterior paintwork. Other distinguishing features included the slightly higher roofline which had been raised from 51inches to 54inches, and the blue police lamp which was virtually standard on all Royal cars. Inside the vehicle, dark blue cloth was used to finish the rear compartment, whilst blue leather was employed in the front.

Queen Mary's car was built on chassis no. 51101, and carried the Hooper body no. 9424 (drawing 8160). Again, it was finished in the standard Royal colours and also had increased interior headroom - in this case, 57inches instead of the more usual 54 inches seen on post-war cars.

On 2nd February 1948, HRH Princess Elizabeth was presented with an attractive Hooper-bodied DE27 limousine. This car, chassis 51040, was originally retained by the Daimler factory after it had been completed in March 1947. However, it was decided by the RAF and the WAAF that they would buy the car from the Company with the money that they had collected to give the Princess a suitable wedding present. Thus, it was acquired for the Royal couple and fitted with the number plate "HRH 1". The standard Hooper six-light coachwork was finished in black, with an interior of blue leather in the front, and beige West of England

Right: *A Daimler DC27 ambulance dating from 1950.*

Below: *A Lanchester LD10 rolling chassis.*

cloth in the back. The windows were all electrically operated, and there was also a radio. The Royal Crown was painted discreetly on the two rear doors and a mounting point for the Royal Standard was provided, situated just behind the customary blue police light.

The 1948 Motor Show

The year 1948 will always be best remembered in motoring circles for the first British post-war Motor Show, held at Earls Court in October. At the show, Daimler was well represented, not only on the Company's own stand, but also on several of the coachbuilders stands, too.

Back in the autumn of 1948, Daimler had introduced their Special Sports chassis, which was a higher powered version of the proven DB18. The car was designed for effortless high-speed cruising, and its power unit and layout owed a great deal to its predecessor, the Daimler Dolphin. The Special Sports engine produced 85bhp, compared to the 90bhp of the Dolphin unit, and it is easy to see why this new 2.5litre chassis became so popular with the Press and public alike. A standard Daimler transmission was used, which incorporated an overdrive epicyclic gearbox.

The standard post-war DB18 saloons had retained their now rather dated vertical radiators, but the fluted grille of the Special Sports was given much the same treatment as the larger DE27s and Straight-Eights. The vehicles body weight had been kept down to a minimum, and aluminium was used on all the panels except for the front wings, which were steel. The folding hood was designed in such a way that external landau irons were omitted, and spring assistance helped to ease its operation. Although the hood did not disappear completely when lowered, it did at least lie flatter than the vast majority of similar folding hoods.

The bench seat in the front of the car could comfortably accommodate two people, or three at a push, and the leather-covered front seat squabs would tilt separately to allow access to the back. In the back of the vehicle there was a single seat mounted sideways, and although this was not an entirely new idea, it was certainly unusual. When not in use, the seat could be folded away, giving more room for the larger items of luggage. The Special Sports featured a full set of instruments, and polished walnut was used for all of the woodwork.

A Barker Special Sports (at least 75% of Special

Left:
Lanchester LD10s received Barker bodies after June 1950. Some 580 Barker bodied cars had been built when production was brought to an end in the following year.

Below:
Announced in September 1949, the Consort was the long awaited replacement for the highly successful DB18 saloon.

Sports chassis had Barker bodies), chassis no. 53750, finished in red and cream, was awarded the Silver Medal in its coachwork class at the Earls Court Show. Glamorous, the Special Sports, in the words of *Autocar*, *"inevitably evokes a picture of the blue of a Mediterranean sky and sea as the perfect setting for it after a high-speed run down the N6 and N7 to which its effortless fast cruising is so suitable."* At around £2,000, the Jim Rogers designed Barker Special Sports was built to appeal to this sort of market.

The Windovers stand proudly displayed a DE27 with a type 136 saloon body. This car (chassis no. 51261) had no division, but featured an optional separate adjustable driver's seat with a fixed bench seat alongside it. The coachwork was finished in black and maroon, with the leather interior also in maroon. The price for this particular car was put at £3,267 including purchase tax.

Lancefield Coachworks of London exhibited an unusual Daimler at both the 1948 and 1949 Motor Shows - these were special invalid limousines built on the DE36 Straight-Eight chassis. Similar cars had been shown before the war, and although the post-war models were not particularly attractive to look at, the judges at the 1948 Show thought highly enough of the vehicle to award its makers a Silver Gilt medal. The body featured three full length doors with drop windows, a concealed luggage rail on the roof, and plenty of luggage accommodation at the rear of the car. Inside was a complete set of medical equipment, a full length stretcher mounted on its own chassis, and a luxurious revolving chair which was housed in the division when not in use. The division itself was electrically operated, as were the rear window blinds; a radio and separate heater were also fitted in the rear compartment. However good the idea may have been, only two, or possibly three, of these limousines were ever built.

Hooper's did well in the Institute of British Carriage & Automobile Manufacturers coachwork competition, taking first place with their superb Daimler Straight-Eight touring limousine. The vehicle (chassis no. 51256) was finished in mistletoe green, and was trimmed in a light beige leather. The price for the complete car was just over £6,500 with purchase tax.

However, the Green Goddess, as the car was dubbed, was undoubtedly the star of the show. With its streamlined drophead coupé body by Hooper, this gigantic machine was, at about £7,000, probably the most expensive car there. Based on the DE36 chassis

Right:
An interesting exhibit at the 1949 Motor Show was this magnificent Straight-Eight sedanca de ville by Hooper.

Below:
Lady Docker owned this delightful sports coupé built by Hoopers on the 2.5 litre Special series chassis.

(no. 51233), the Green Goddess was of immense size, a full 20 feet long and 6ft. 5ins. wide. The body shape was quite unlike anything seen on a Daimler before - the front wing line was extended so that it tapered out at the rear wheels, and the rear wheels themselves were covered by aluminium panels fitted with spring balance arms. The rear end of the car was rounded off in one clean, if somewhat heavy sweep - overall the car was very pleasing to look at.

The two gigantic doors were fitted with electric windows, and the front windscreen was not only very wide, but also curved. The Green Goddess had three front seats, all with separate hinged backs to allow for easy access into the rear of the car where there were two more seats. The electro-hydraulically operated hood was of particular interest, as it folded away completely under a power-operated metal cover positioned just forward of the large boot. The traditional Daimler fluting was carried over to the tops of the stylised headlamp covers.

Although an official road test was never carried out on the Green Goddess, it is understood that the car was capable of speeds in the region of 90mph. After a lot of development work, eventually around eight cars were built along the same lines, with only slight differences in their detailing.

Special Straight-Eights

Another interesting car based on the DE36 chassis, but this time with coachwork by Freestone & Webb, was completed in early 1949. It was built for Ivan Hendricks, who lived in Jamaica at the time, but had worked for Daimler during the war years. Although the car was finished in 1949, it had taken nearly two years for the project to become the beautifully proportioned vehicle shown in the illustration. The graceful lines belie the fact that the car was a full six-seater, and although the price was never publicly disclosed, this must have been a very expensive car indeed. The paintwork was finished in a soft pastel grey with a thin gold coachline. A slightly darker shade of grey was used for the leather interior, and all cabinet work was finished in a special

Left:
One of the extremely rare Hooper bodied Special Sports chassis. This example, owned by David and Marie Adcock, was reputedly built for King Farouk of Egypt, although it was not delivered to him since he was deposed before it was finished.

Jamaican wood known as yellow sanders. The controls on the fascia and steering wheel were ivory, and in the rear of the car, folding tables were mounted on either side of a central cabinet.

The chassis (no. 51238), was left almost completely standard, except that all of the metal parts under the bonnet were highly polished, and the steering column was lowered by about 4inches - the suspension was also altered slightly to allow for the change in the cars weight distribution. In fact, a sister car was built, which was completed just in time for it to be exhibited at the 1949 Earls Court Show. This car was finished in beige with the wings in a contrasting deep maroon; the seats and door panels were trimmed in beige hide with a maroon piping, and all interior woodwork was of highly lacquered burr walnut. This car was some 18ft. 8ins. long, 6ft 2.5ins wide, but only 5ft 5.5ins high.

Two more State cars were delivered to King George VI on 7th June 1949 - they were to be his last State Daimlers before his death in February 1952. The two pre-war Straight-Eight landaulettes were taken by Stratstones in part exchange for the new cars, which were of an identical design to those supplied in 1947.

On delivery, it was declared that chassis no. 51741 (body no. 9542) should become the new No.1 State car, and that chassis no. 51740 (body no. 9541), would be the No.2 car. The 1947 examples were thus relegated so that the old No.1 car, which had been used the most, became No.4, and chassis no. 51172 was consequently graded as No.3.

Another Royal Tour

A Royal Tour of Australia and New Zealand was planned for 1949, but was cancelled due to the King's poor health - in fact, the Tour did not take place until the end of 1953, after the Coronation of Elizabeth II. However, the cars had already been ordered by the respective Government's.

For this Tour, Hoopers had been asked to produce bodies for twelve DE36 Straight-Eight chassis, which consisted of four all-weather cars, five landaulettes and three limousines. Many lessons had been learned during the South African Tour, and the findings were put to practical use for the new cars. Coachwork was again finished in blue and black with a fine red line; the all-weather cars were trimmed in red leather with a fawn hood, and the closed cars and landaulettes had blue upholstery in the front, with grey cloth to the rear.

Bob Crouch had again been chosen to represent the Company, and to supervise the care of the cars. All of the vehicles had been inspected by the King and Queen on 29th October 1948, the trade and Press day had been and gone, and the cars had even been driven down to the docks and made ready for their long journey. Everything was ready, when at the very last minute, the news came of the Tour's cancellation. The DB18 saloons, which had been displayed at the Hooper showrooms before the Tour, were sold through the usual trade channels, but the larger specially built DE36s, were kept until just before the Christmas of 1953, despite an article stating six of the cars were for sale in early 1949.

Variations on the DE27

During 1949, two interesting variations on the DE27 were built, the first of which being designated the DC27. In most respects, the two chassis were identical, though this new version was developed for use solely as an ambulance. The only significant modification concerned the transmission layout, which featured two propshafts mounted in series, the forward one being supported by a rubber-mounted centre bearing. The hypoid bevel rear axle and differential unit were positioned slightly to the nearside of the vehicle, and the engine was also arranged differently. Not only was it inclined backwards, as seen on the standard DE27 chassis, but also very slightly diagonally to suit the new transmission set-up.

The engine itself featured a single Solex downdraught carburettor in place of the twin SUs seen on the DE27. This had the effect of marginally reducing the power to around 105bhp, but the Daimler ambulance was indeed a fine machine, and was produced in several different forms by both Barkers and Hoopers. In all, 500 DC27s were built, with the chassis numbers running from 54000 to 54499.

The other variation on the Twenty-Seven chassis was the DH27; a vehicle specifically designed for use within the Daimler Hire outfit - part of the Thomas Tillings Group since 1931. The DH27 was actually a development of the DC ambulance chassis, and there were no significant changes necessary. The body was that of a large limousine built by Hoopers, and about 50 were produced in total (chassis D 52900 to D 52949). With an overall length of some 18ft. 9ins. and a width of 6ft. 6ins., it is probable that these were the largest hire cars ever built in quantity. There was ample accommodation for up to seven passengers and a chauffeur. Air conditioning, efficient heaters, electrically operated partitions, and large 'sight-seeing' windows were all standard features.

During the same period, a fleet of specially constructed tandem drive six wheeled double-deckers were supplied to a company in Cape Town, South Africa. By the early 1950s, the Daimler double-deckers were of a far lighter construction than before, and soon all the front engined commercials were sporting a new style grille in place of the familiar fluted radiator surrounds.

Below: *The highly desirable Hooper Empress, circa 1950.*

Left:
Exhibited at the Earls Court Motor Show in 1950, this DE36 with a sedanca body by Hooper gained joint first place in Section I of the coachwork competition.

THE "SPEY" CLOSE-COUPLED OR TOURING LIMOUSINE BY HOOPER

Right:
This Hooper close coupled, or touring limousine was christened the "Spey". Not all of them had enclosed rear wheels.

Lanchester news

On the Lanchester side of things, the LD10 had been given a totally new Barker body. Constant delays in the delivery of pressed steel body panels led to a stoppage in production during November 1949, and no more LD10s were built using the original Briggs coachwork. As noted earlier in the chapter, 25 LD10 prototypes were built, and these carried the chassis numbers 60000 - 60024 inclusive. The first production model (chassis no. 60025) was built in April 1946 - 1,000 cars followed, with consecutive chassis numbers, the last of which being produced during June 1947.

However, for the next batch of cars (built between June 1947 and November 1949, and still with Briggs bodies), a new batch of numbers were allocated; 61800 - 63225. The next run of LD10s, with their more elegant and partly aluminium alloy bodies, went into production in June 1950, but unfortunately the model had been priced out of the market. At £725 plus purchase tax of £445, only 579 Barker Tens (chassis no's 63226 - 63804) were sold. The Barker-bodied Lanchester LD10s were produced for about a year, before the Fourteen came along to replace the model.

The Daimler Consort

The Consort, which had been announced in September 1949, was finally put into full scale production during the early part of the following year as a replacement for the now somewhat dated DB18. It had originally been marketed for export only, and for this reason was built concurrently with the standard DB18 saloon, the dropheads, and the Special Sports.

The Consort, as a development of the DB18, had many of its predecessors features. The engine had been left much the same; it still produced 70bhp, but the pistons were now fitted with three compression rings. Girling hydro-mechanical brakes replaced the DB18s mechanical ones, and wider wheel rims were employed to aid braking, handling and traction. Dr F.W. Lanchester's worm final drive, which had been a feature of all Daimler cars since 1910, was superseded by a hypoid bevel gear on the Consort. The larger

Right:
This DE27 with six light limousine coachwork by Hooper, was built for the King of Siam in 1950. Note the chrome slats along the bonnet side - usually a distinguishing feature on the larger Straight-Eight models. Just over 200 Twenty-Sevens were finally produced.

Below:
The Lanchester Fourteen, or Leda as it became known.

DE27s and DE36s were already equipped with this type of rear axle of course, leaving only the Special Sports to continue using the worm. The overall length of the Consort was some 1.5ft. greater than that of the old 2.5litre saloon, and consequently, its turning circle was also slightly increased.

The responsibility for the all steel six-light coachwork was left in the capable hands of Mulliners, and was in fact quite similar to that of the standard DB18 saloon. However, there was a new bowed radiator of the same design as that used on the Special Sports (although the vertical grilles were painted to match the body colour on the new car, instead of being chromed); the headlamps were faired into the detachable wings, and the rear end treatment was rather more pronounced. Other differences included new disc type wheels with larger hub caps, new rounded bumpers, and an enclosed rear number plate. Sunroofs were no longer fitted as standard to the Consort, but there were now hinged quarter-lights on both front doors. Inside the car, the separate seats were arranged in such a way that three abreast accommodation was just possible, and armrests were provided. There were only slight detail changes to the fascia, and an adjustable steering column was now a standard feature.

The maximum speed of the Consort was said to be just under 80mph, with fuel consumption averaging around 17mpg to 20mpg. The cost of the vehicle varied quite considerably through its fairly short life, from around £1,600, to over £2,000 at one point. A few Consort chassis were sold separately, and some of them were mounted with Barker four-seater drophead coupé bodies similar to those of the original Barker and Tickford dropheads, which had been discontinued along with the old DB18 saloon in 1950.

One Consort, driven by A.E. and A.W.M. Adams, is known to have entered the 1951 Monte Carlo Rally, and won its Class in the Concours de Confort (the event was won for a fourth time by Jean Trevoux, this time in a Delahaye). A number of privateers actually used the Daimler to good effect in the more gentle days of rallying.

Enter Lady Docker

Sir Bernard Docker was married on the 3rd February 1949 to Norah Collins (née Turner). Born in Derby in 1906, and having served in the Mechanised Transport Corps during the war, although she was Docker's wife (at least after the private ceremony in Caxton Hall, Westminster), she was to have a great bearing on the future of the Daimler Company, in more ways than one.

Once briefly married before to an actress, Lady Docker said that Sir Bernard's best feature in the early days was his 863 ton yacht Shemara - it was her third marriage, and all of her partners had been millionaires! However, they grew together and before long, Lady Docker had noticed that the Daimler marque was *"in danger of becoming a relic."* She is said to have told Sir Bernard, *"The only people who know about the Daimler are the Royal Family! I know you couldn't find better customers, Bernard, but Daimler can't survive on status alone. It's got to sell to the masses."*

Left:
The Ferret Mk. II.

Right:
A Daimler type CTC6 trolleybus dating from around 1950. This particular bus was later exported to Spain.

Left:
A typical horse box of the period mounted on the popular Daimler commercial chassis.

In keeping with the family tradition *("jobs for the boy's")*, Sir Bernard found a job for Richard Smith, Lady Docker's brother-in-law. He was given a management post at Daimler and, in 1954, following a take-over, a seat on the Board of Carbodies. Following her new found interest in wanting to put Daimler back on the map, he also made Lady Docker a Director of Hoopers.

A special Special Sports

One of the few special bodied post-war DB18s was an unusual two door fixed head coupé, built by Hoopers on the 2.5litre Special Sports chassis for Lady Docker. Mechanically, the car was a standard Special Sports chassis (no. 53790) with its twin carburettor engine producing some 85bhp. The body (no. 9579, and completed in June 1950) could be described as a two door version of the famous Empress, first shown at the Earls Court Motor Show in 1950.

The main alterations were concentrated around the front of the car. The head, side and fog lights were all enclosed beneath a perspex shield which followed the contours of the wing. A chrome surround finished the feature, the upper part of which being fluted in the style of the Green Goddess. The interior had been designed with touring in mind, and the rear compartment contained two large and comfortable individual seats, both of which could be folded back to provide more luggage space.

There were also six special drophead coupé versions of the DB18 were built by Hooper, based on the Special Sports chassis. The hood was power operated and when lowered, a metal lid situated behind the rear seats opened electrically, before closing again to completely conceal the hood - this was another feature which Hoopers had carried over from the 1948 Green Goddess. The first of these cars (chassis no. 53806) was delivered to King George VI in July 1950. The other five were built between March 1951 and August 1952, and carried the chassis numbers 53890, 53898 (which was delivered to ex-Queen Marie of Romania), 53903, 59113 and 59114.

Something completely different from 1950 was the so-called Reville Ranger. Jean Reville had once built a limited number of fwd BSA based sports cars in London, but had then moved out to Australia. From there, he sent news of his intention to build a new motor car - the Ranger - using a Lanchester Ten engine and fluid-flywheel transmission, combined with the Reville multi-variable gearbox. The top speed was estimated to be in the region of 70mph, but sadly nothing more is known about the project.

At around this time, the Company had obtained a post-war Invicta Black Prince engine and gearbox. Apart from the technology associated with the Meadows dohc power unit, it was most likely the Brockhouse transmission that was of interest to the Daimler engineers, as it allowed an infinite variety of ratios to be selected automatically. These may have been obtained through Charlesworth, but following Invicta's closure, Frazer-Nash (who bought the assets) collected the various components from Coventry. The Invicta marque was never revived, and Frazer-Nash failed to make any use of either the engine or the gearbox.

The 1950 Motor Show

At the 1950 Earls Court Motor Show, the Daimler Company was again well represented. The Empress, a supremely elegant body style produced by Hooper & Co., was first mounted on the Special Sports chassis, but continued to be seen in various forms on the Daimler models until the latter half of 1958. Some Empress models had built in fog and spotlights and a one-piece bonnet hinged at the rear, while others had air intakes at the front and centre-hinged bonnets; there were also slight differences in fascia layouts. The Empress for the 1950 Show (chassis no. 53811) was finished in two shades of green with a pale green leather interior, and had a number of special features - perhaps the car's single most important feature was its low overall height. Other features included the double sealing of all the doors, and the roof was specially insulated against heat.

The DE36 on the Hooper stand also did well in the coachwork competition, taking joint first prize in its class. The Sedanca body (no. 9620 on chassis no. 52823) was finished in cream over black, and was styled

Below:
Daimler Straight-Eight chassis no. 52830, better known as the Gold Car - the 1951 Earls Court Motor Show exhibit.

along the same lines as the Green Goddess, but with a different top half. White cloth was employed to trim the rear compartment, whilst in the front, black leather was much in evidence. No occasional seats were fitted, but the car could seat five passengers comfortably, in addition to the driver. The de ville extension (which was patented by Hoopers) was a notable feature, and so too was the cocktail cabinet which would move back and upwards towards the rear occupants at the touch of a button; the door then opened to form a table, and a light inside the cabinet illuminated the contents. Sadly, this important car was later converted into a limousine.

The DE27 made its final appearance at Earls Court during the 1950 Show. Its production came to an end long before the 1951 event, the last one leaving the works in the July of that year.

A new Lanchester - the Fourteen - was announced in October 1950, destined to replace the LD10. Full scale production of the Lanchester Fourteen began during the early part of 1951. In many respects, this new Lanchester was an important model; the engine was developed into a three litre six cylinder unit which powered the Regency, and the body and chassis were the basis for the best selling Daimler Conquest, which made its debut in 1953.

Introduced at a price of £895 (plus £496 purchase tax), the Fourteen had a revised chassis, and a brand new four cylinder two litre engine. It developed 60bhp (which was not really powerful enough for the body of the car) and had a bore and stroke of 76.2mm x 108mm, giving a total displacement of 1,968cc. The revised chassis (designated type LJ200 for right hand drive models, LJ201 for left hand drive cars) in fact bore a striking resemblance to that of the pre-war LA14, except that the coil springs used in the front suspension of the LA14 were replaced by torsion bars and wishbone links in the Fourteen.

1951

The Daimler Freeline single-decker coach made its debut in 1951, and had its engine (either a Daimler or Gardner unit) fitted underneath the floor. This model proved to be quite successful, and sold in fairly large numbers not only on the home market, but abroad as well.

The Company once again had Armoured Cars listed as one of its staple products following the introduction of the Rolls-Royce B60-powered Ferret. The Ferret continued in production well into the 1970s, albeit in an updated form, and was used by the NATO forces as one of their standard wheeled vehicles for a number of years. Dennis Jenkinson, the famous motoring writer, once described the Ferret as *"diabolical"*, but considering he was more used to driving thoroughbred sports cars, it is hardly surprising!

Lady Docker's Magic

For some time, Lady Docker had been working with Osmond Rivers of Hooper to produce something that

Left:
The Daimler Regency in a contemporary setting.

would *"make Daimler a household name, and to keep it in the minds of the vast car-buying public."* At the 1951 Earls Court Motor Show, a Hooper Empress once again took first prize in the coachwork competition. Carrying chassis no. 56183, it was identical to the previous year's car, except for its maroon/cream exterior and cream leather interior. The award for the most looked-at car of the event, however, must have gone to another vehicle on the Hooper stand, the "Gold Car".

The Gold Car was the first in a famous line of splendid show cars that would make the Daimler marque something of a legend. The car was built for Sir Bernard and Lady Docker on the DE36 chassis (no. 52830) at a cost of £8,500 - a very large sum in the austere days of the Clement Attlee Labour Government. The huge, but well proportioned touring limousine, was finished almost entirely in black, except for the side panels below the waistline which were black with 7,000 tiny gold stars. Every part on the car which would normally have been chrome plated had been dipped in real gold, from the largest component right down to the smallest detail, according to Lady Docker, because *"it was practically impossible to obtain chrome"* and brass would tarnish.

The upholstery was trimmed in golden silk, as was the headlining, and both had to be specially woven on a handloom. Hoopers themselves carried out the cabinet work, which was built using golden Australian camphor wood, and Cartier of Bond Street supplied the gold and crystal picnic fittings. The china tea service was in black and gold to match the rest of the car, while the drinks and so on were kept in the opposite cabinet in the division. There were also black suitcases made of crocodile hide, design specifically for the car's large boot.

Over the rear compartment, there was a toughened glass roof panel with an electrically controlled shutter. The division and all the side windows were electric, in spite of the latter being double glazed. There were separate heater units for the front and rear compartments, and the rear window was heated internally to prevent misting - an innovative feature well ahead of its time.

Several of the features found on the Green Goddess were carried over on the new car, such as the perspex headlamp covers and the flush fitting rear wheel covers mounted on spring balanced arms. In fact, there was strong family resemblance between the two cars.

The Gold Car was one of a batch of ten or so selected vehicles on which Hooper modified the front suspension. It was found that by replacing the original front shock absorbers with telescopic dampers, and mounting the radiator differently, it was possible to isolate both the radiator and the front wings from the rest of the chassis, thus eliminating any vibrations or shaking. It goes without saying, but the Gold Car was rightfully given the premier award for the event.

It is sad to think, that after just five or six years of service, the Gold Car, like the other Docker Daimlers which followed it, was stripped of all its special features and sold after a few suitable modifications. In the case of the Gold Car, the front was remodelled, all the gold was removed, and the body was resprayed in fresh black paint.

A New Daimler

The Gold Car had overshadowed the production Daimlers at the 1951 Motor Show despite the fact that one of them was an important new model. The Regency, as it was known, had been shown at the Paris Show a few days earlier, but this was the car's debut in the United Kingdom. At this time of course, the 2.5litre Daimler Consort was still in production, although the Regency represented quite an advance over the older car. It was to be the first in a new series of modern Daimlers, and used a six cylinder version of the proven Lanchester Fourteen power unit.

The Regency engine used the same bore and stroke dimensions as the Fourteen (76.2mm x 108mm), but with six cylinders, the capacity of the unit was increased from two litres to 2,952cc. Following established Daimler practice, the overhead valves were operated by pushrods, and were slightly inclined - in this case set at about 20 degrees. The combustion chambers were still of the familiar lozenge shape, despite having been slightly modified. A considerable amount of work had gone into the design of the new cylinder head in a bid to avoid distortion. Ample cooling was provided around the top of the block, the exhaust ports and spark plug bosses, while the head bolts were located well away from the bores.

Other features aimed at giving the engine a long life were applied to the rest of the unit, such as the use of special Brivadium cylinder liners. A four bearing crankshaft was employed, and the ignition system incorporated an unusual control which enabled the driver to adjust the advance and retard according to the grade of fuel he was using. As on the earlier models, the induction manifold was again water-jacketed, although there was a new arrangement for pressure balancing the twin SU carburettors. Maximum power was put at

Right:
Viewed from above, the new convertible looks slightly heavier around the rear wings than the Special Sports did, although it is, nonetheless, every bit as pretty.

Below:
The Regency Coupé - a worthy successor to the Special Sports series of cars.

90bhp at 4,100rpm.

The chassis (designated DF300 for right hand drive cars, DF301 for left hand drive), was a development of the Consort frame, although only the wheelbase remained the same dimension on both models. With the robust new chassis, ground clearance was increased from 6inches to 7inches, even though the side-members had been lowered slightly.

To accommodate the larger three litre power unit (and to provide more room within the car), the engine and gearbox were moved well forward, and following the trend set by the more recent Daimlers, the Regency was fitted with a hypoid bevel rear axle. The front suspension, similar to that employed on the Consort, consisted of independent coil springs with single upper and lower transverse links of unequal length on each side, supplemented by a radius arm mounted on the chassis. There was one notable difference, however, and that was the use of telescopic dampers fitted inside the coil springs - an anti-roll bar was also employed. The rear suspension made use of semi-elliptic springs in conjunction with telescopic dampers. The braking system was still of the Girling hydro-mechanical type, though the friction lining area was considerably increased for the new model.

The Regency body (built at the Radford Works under the Barker name) was designed to offer increased space and comfort for passengers, and was very modern by Daimler standards. The interior could easily accommodate three people in the front and a further three in the back thanks to the use of fashionable beach seats - armrests were provided on all four doors and in the centre of the rear seat. The fascia was of polished walnut and the comprehensive set of instruments was grouped in the centre. Provision was made for a radio and a pair of speakers, although these were sold as optional items; twin fog lamps mounted just below the headlamps were a standard feature, however.

Sadly, the Regency was introduced too late. How could it possibly compete on level terms with cars like the Jaguar Mk.VII, introduced in the previous year? The price of the new Regency model was put at £1,500 basic (or £2,334 16s 8d including purchase tax), whereas the Jaguar saloon - which was quicker, handled better, and was every bit as luxurious - was less than the basic price of the Daimler including tax.

Corporate matters

Some wheeling and dealing between some of the larger Midlands companies had gone on during 1951. BSA acquired Triumph motorcycles for around £2.5 million, securing the services of Jack Sangster and Edward Turner (their dictatorial Managing Director), into the bargain. Jaguar, desperate for extra capacity, managed to obtain a lease on the Daimler No. 2 Shadow Factory at Allesley. About a year later, on 28th November, the site at Browns Lane was officially declared Jaguar's.

More new cars in '52

During June 1952, Appleyard's of Leeds announced that they could convert the Special Sports drophead coupé into an attractive fixed head model. The body of the car was left completely standard, with only the removal of the fabric roof and its framework being deemed necessary. A new frame was built using ash, and this was subsequently panelled in aluminium which in turn was welded to the body at the back, and lap jointed at the base of the windscreen pillars at the front. The rear window and quarter lights had to be designed specially to suit the conversion, which was so pleasing it gained the full approval of the Daimler Motor Company itself.

At the Paris Show in October 1952, the Company displayed a superb 3litre Regency convertible coupé. The coachbuilt body was credited to Barker & Co., and was closely based on the old but highly-successful 2.5litre Special Sports.

The hood was power operated, as were the windows, and even the boot lid was balanced in such a way that it rose automatically once the handle had been turned. The brochure stated that the new model had been designed with high speed long distance travel in mind, so the two door body was fitted with two comfortable bucket seats and a centre armrest in the front, and a full width bench seat for the back, again with a central armrest. To enable the car to obtain high speeds and sustain them, a new aluminium head was fitted to the engine, and the compression ratio was raised to 7.5:1. This had the desired effect of increasing the power output to around 100bhp at 4,400rpm. In addition, the epicyclic gearbox fitted to the Regency convertible model was given an overdrive top ratio for effortless cruising, just like its 2.5litre predecessor.

By utilising twin fuel tanks, one in each of the rear wings, it was found that the space this provided under the boot could be used to store the spare wheel. There was nothing particularly new about the arrangement, but Daimler decided to mount the spare wheel in a retractable cradle underneath the floor. By using the cars wheel brace, this could be lowered to the ground, eliminating the need to unload the boot, which is inevitably full in the event of a puncture. Unfortunately, no road test report was ever compiled on the Regency Convertible Coupé (designated DF302/3), although the price for this highly-regarded machine was put at £1,710 basic, or £2,661 with purchase tax.

In fact, very few three litre Regency chassis were ever built - without doubt, less than 50 of them, and the majority of these were DF302s which found their way to Hoopers to be fitted with special coachbuilt bodies. One Regency saloon (DF300) was retained by Daimler and used for many years as the works hack in Coventry. Later in its life, it was converted into a pick-up truck, and was fitted with a replacement 3.5litre engine.

Earls Court 1952

The superb Hooper Empress had proved so popular on the 2.5litre Special series chassis, that the famous coachbuilders continued to build the model on the DF302/3 chassis. The coachwork had to be made slightly larger to conform to the new chassis dimensions,

Below: *A magnificent Straight-Eight Daimler - a special fixed head coupé built for the 1952 Motor Show.*

Left:
This rather ugly DE36 was bodied by Saoutchik et Cie of France to the order of HM King Ibn Saud in 1952.

but this in fact made the vehicle even more elegant, as it had now become a well proportioned six-light sports saloon.

One such car (chassis no. 82002) painted in two shades of green and trimmed in grey-green leather throughout, was exhibited by Hooper at the 1952 Earls Court Show. The car was highly commended by the judges, who awarded the vehicle the Institute's Silver Medal. Of the 33 Mk.II Empresses built on the DF302/3 Special series chassis, 31 were standard four door models, with the remaining two being of a specially-constructed two door design.

Another Docker Daimler made its first public appearance at Earls Court in 1952; the "Blue Clover" sports model. Built on the legendary DE36 Straight-Eight chassis, Blue Clover was a rather attractive car produced by Hoopers to the order of Sir Bernard Docker. Designed primarily for long distance touring, a considerable amount of luggage space was made available, with the seating also being arranged to give maximum comfort.

The body design was a fairly simple one, its graceful lines making the car look very much like a refined fixed head coupé version of the original Green Goddess. The coachwork was finished in tones of powder blue and grey, with the lighter panels being covered with a uniform pattern of four leaf clovers in the darker shade of grey.

Instead of using wood veneers for the interior furnishings, it was decided that grey-blue lizard skins should be employed. The steering wheel cover, door finishers, fascia board, and the cabinets and armrests which flanked the armchair style rear seats (which could be folded to give extra luggage space) were all trimmed in this exotic material.

All side windows were electrically operated and a transparent panel was provided in the front part of the roof. Double glazing was once again used to keep out the heat, cold and noise, and thermal insulation was fitted in the roof. All in all, Blue Clover was a highly desirable motor car, and indeed, one contemporary writer stated, *"it is the most elegant thing at Earls Court."*

Earlier in the year, a Straight-Eight had been entered on the 1952 Monte Carlo Rally. For rallying, a coachbuilt Hooper limousine was rather an odd choice, but the car (said to be *"very sure-footed for so elephantine a vehicle..."*), driven by Tommy Wisdom, Lord Selsdon and Anthony Hume, in fact performed very well indeed. The team followed the Lisbon route, and although they were actually unplaced at the end of the rally, both the Daimler and the crew managed to finish the course with honour - one of only 167 finishers from 369 starters.

A new direction

The famous Saoutchik bodied car for King Ibn Saud, the DE36 which took part in the Monte Carlo Rally, and Blue Clover were among the last of the more notable Straight-Eights to be built. Sadly, production

Left:
A Lanchester Fourteen on the 1953 Monte Carlo Rally.

of the DE36 came to an end during the November of 1953, marking the end of a magnificent era.

Two other established Daimler models were also discontinued during 1953 - the DB18 Consort, and the 85bhp Special Sports (the last car to feature the Lanchester worm-drive rear axle). These changes were doubtless made in the name of progress, but the new line up of vehicles became generally less rarefied, and this may well have had a damaging effect on the marque's prestige. However, the car that in effect replaced the Consort - the Daimler Conquest, became a very popular model, and in fact went on to be the Company's best selling machine of all time.

Introduced in May 1953, the Conquest was basically a Lanchester Fourteen (by now referred to as the Leda after a few minor modifications), but with a larger six cylinder engine. The chassis frame was almost identical to that of the Leda, and most of the body pressings were exactly the same. With this in mind, contemporary reports which state that the Conquest went from drawing board to production in less than four months could well be correct. The Conquest was aimed at those who required a high standard of finish, combined with accommodation for the family and respectable performance, without the costs associated with running a larger car.

The answer to why the Conquest was so successful compared to the Lanchester version, lies with the engine. It was a typical Daimler unit, the bore measurement being the same diameter as that of the 2litre Leda and the 3litre Regency (76.2mm), and the stroke set at 88.9mm to give a capacity of 2,433cc. The block was produced in cast-iron with dry liners and supported a crankshaft with four main bearings. The camshaft also had four main bearings, and was made in chilled cast iron. At the bottom of the engine, the oil pump and crank chamber were enclosed by a deep sump, cast in a light alloy.

The valve gear and cylinder head itself, although fairly conventional, had several interesting features. Special attention was given to improved gas flow, and once again the valves and porting were slightly modified. The valves were inclined at 7 degrees (10 degrees had been used on the Consort) and the inlets were of a larger diameter than the exhaust valves. A lozenge-shaped combustion chamber was still found to be the most suitable, and the long reach 14mm spark plugs were fitted at an angle on the nearside of the head. A single downdraught Zenith carburettor was employed, and the aluminium alloy inlet manifold featured separate porting and water jacketing, as seen on the Regency engine.

With this new engine, particular attention was paid in eliminating unnecessary weight, and in all, about 1cwt was saved when comparing the Conquest power unit to that of the Consort. Another piece of

Left:
The aggressive lines of the Regency fixed head coupé, pictured here in January 1953.

Left:
It would be difficult to imagine anything less suitable for post war rallying than a Straight-Eight Daimler, but this 1951 car put in an admirable performance on the Monte Carlo Rally. It was driven by Tommy Wisdom, Lord Selsdon, and Anthony Hume.

Below:
The interior of the three litre car.

Below:
Another view of the rare DF303 Regency fixed head coupé - probably one of only four built.

thoughtful design concerned the inclusion of a special hole in the sump which enabled an external oil feed to be fitted when the engine was on the testbed. After initial running in, the unit could be thoroughly flushed, and the hole sealed with a plug.

Maximum power was put at 75bhp (at 4,000 rpm) with a compression ratio of 6.6:1, although this was later changed to 7:1. Fluid flywheel transmission and an epicyclic gearbox were employed to good effect, with a Hardy-Spicer open propshaft taking the drive to a Salisbury hypoid bevel rear axle, normally with a ratio of 4.56:1.

The chassis frame was a conventional sturdy design, with box section longitudinals being linked by channel section cruciform bracing and massive crossmembers to the front and rear. The front crossmember passed beneath the forward end of the crankcase, and was made to carry the front suspension. This consisted of laminated torsion bars and forged wishbones of unequal length - a layout first seen on the post-war Lanchester Fourteen (or Leda). Long semi-elliptic springs were used at the rear, and Girling telescopic dampers were fitted all round.

Braking was via machined and drilled drums on all four wheels. It had been found on some Daimler models, that by drilling holes in the drums, the brakes could be kept up to 50% cooler. As a result, each drum on the Conquest had eight ventilation holes of 1.125inches diameter.

The pressed steel body, as mentioned earlier, was almost identical to a Lanchester Leda, but with a traditional Daimler grille. Inside the car, walnut veneer and leather was used throughout. The front seats were actually separate, although they were positioned so closely that when the occasion warranted, six people could be carried - three in the front, and three on the rear bench seat. A notable feature which aided this arrangement was the steering column-mounted gearchange to the driver's right, and the lack of a transmission tunnel - this was fairly common on all the Daimlers from this period.

The Conquest's maximum speed was about 80mph, and it could cruise quite happily at 70mph. Average fuel consumption was around 20mpg, which for a 2.5litre car at that time was not bad at all. The model was given the chassis designation DJ250 (or DJ251 for left hand drive cars), and the basic price for the Conquest was fittingly set at £1,066. With purchase tax, however, the cost rose to just over £1,511.

The Conquest Roadster (designated Type DJ254/5) was introduced at the 1953 London Motor Show, and was well received by the Press and public alike. A pretty car, aluminium panels over light alloy castings ensured that weight was kept down to a minimum, and the traditional Daimler grille was shortened by some 5inches (when compared to the standard saloon), to suit the machines low overall height.

The Roadster's chassis was a slightly modified version of the type used on the Conquest saloon, but the engine was a little special. Twin horizontal SU carburettors and high-lift camshafts were fitted, as was an aluminium cylinder head, which enabled the compression ratio to be raised to 7.75:1. After the timing was suitably altered, peak power quickly rose to 100bhp at 4,400 rpm (although some of the earliest models had only 92bhp), and speeds in excess of 100mph were easily obtainable. With these changes, fuel consumption improved very slightly despite the increased performance, and the engine weighed some 30lbs less than that of the saloon.

The rear axle ratio on the Roadster was 4.11:1 (a 3.73:1 ratio was available as an option), and Dunlop Road Speed tyres of 6.00 x 15 were fitted as standard. In order to cope with the extra performance, the braking system was upgraded slightly by increasing the width of the brake linings.

The fascia was finished in leather, and in keeping with a sports car, a tachometer was included as standard in the instrumentation. Although in a roadster of any kind, creature comforts are usually of secondary importance, this new Daimler was fitted with a neat hood which folded away behind the rear seats, side screens, and ample luggage space. The car exhibited at Earls Court (built on pre-production chassis 85004) was finished in red, and awarded with the Institute's Silver Medal for coachwork. In the many road test reports that followed, the DJ254 was praised on several occasions for its precise handling and roadholding - not bad at all for a car that was reputed to have been designed and built within just six weeks!

Once production of the Conquest got into its stride, the Company decided to convert two DC27 ambulances to help display the new car up and down the country. Each was fitted with a Brockhouse trailer, and the units toured the principal towns of England with a complete chassis, several components, a Supervisor, and two Daimler Service representatives.

A luxury Lanchester

Another debutante at Earls Court '53 was a rather attractive coachbuilt Lanchester. Known as the Dauphin, this was a two-door saloon with a Hooper

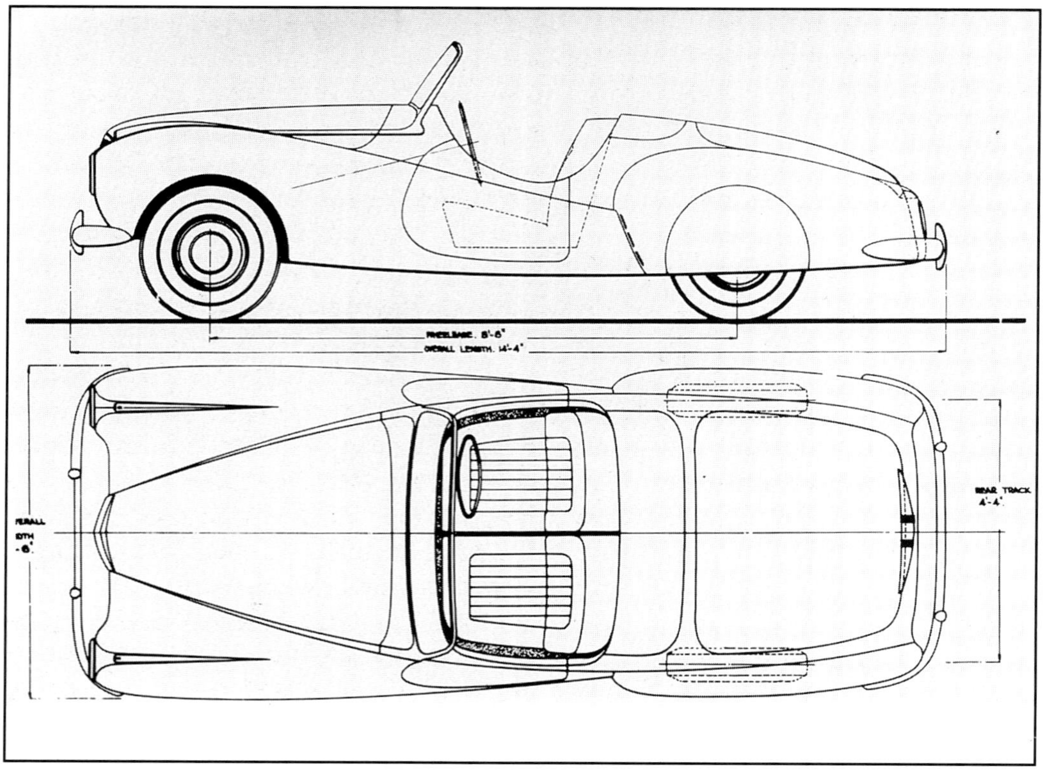

Above:
At this time the Drawing Office was still very active. Here is a proposal dated April 1953, for a neat little Lanchester roadster. If it had been built it would have been some six inches shorter than the Daimler version.

Right:
The Conquest, introduced in May 1953, brought Daimler motoring to a far greater market. Based on the Lanchester Leda, it has been said that the Conquest was put into production in four months.

body and early high performance Conquest 2.5litre running gear. The body was rather like a two door version of the famous Empress line; four seats were provided within the 15ft. 5ins. by 5ft. 7ins. body, and a 92bhp engine was more than enough to allow for swift progress in an aluminium bodied car.

The price of the Dauphin was set a little on the high side at £4,010 5s 10d including tax, and this may have been the reason why only two of these exclusive cars were ever built. A special designation of LJ252 was given to the Dauphins, and the chassis were numbered 70006 and 70007 respectively. One of them (chassis 70007) was destined to reveal its true identity later in life when a Daimler radiator grille replaced its original Lanchester one. This particular car once belonged to R.E. Smith, the General Manager of the Company, and was first registered in March 1954.

"Silver Flash"

By this time, people were expecting something a little special each year from Daimler; the 1953 Show car was "Silver Flash". A car similar to Silver Flash had been built on a prototype 3litre chassis (no. 57004) in the early part of 1952 for Lady Docker, but the Silver Flash itself was constructed on a production DF302 (chassis no. 82001).

Silver Flash was a fairly staid exhibit by Docker Daimler standards, in the form of a two seater fixed head coupé built for Lady Docker, and first painted in dark green. However, shortly before the opening date of the show, it was resprayed in a metallic silver colour, hence the cars name thought up by Lady Docker herself.

Hoopers (Lady Docker was in fact a Director of Hooper & Co.), built the vehicles unusual aluminium alloy coachwork, which was kept as low as possible and comparatively light. The radiator grille was particularly interesting, as it represented a complete departure from tradition, and was in fact similar to that which was fitted to the 2.5litre Jaguar-Daimlers of the mid-1960s.

A Sundym glass roof panel was fitted, and an internal shutter was provided to block out the light when required. Black leather with red piping was used to trim the interior, and red crocodile was used on the fascia board, door cappings, fitted suitcases and so on. Overall, the car was 16ft 8ins long, 6ft wide, and weighed about 36 cwt.

A Tour of the Antipodes

The long awaited Royal Tour of Australia and New Zealand began in December 1953, and continued well into the following year. The Daimler cars which had been ordered for the cancelled Tour of 1949, and which had been gathering dust for so long, could at last be given an airing.

Bob Crouch, who had represented the Company on the South African Tour, and who had been asked to look after the cars during the intended 1949 Tour, was unfortunately engaged in other Company business at the time, so his place was taken by George Fabel of the Experimental Department.

Fabel, who had rallied with Crouch in the pre-war days, had been responsible (under C.M. Simpson) for the Regency series of engines - among other things - and the variants that followed, and was directly involved in the development of the VIP Special series 3litre chassis for HM the Queen's Empress. Born Ronald Fabel in London, he joined Daimler to finish his apprenticeship following Riley going into Receivership in 1938. By 1949, he had been given the post of Chief Experimental Engineer, and continued with the Company until his retirement.

All of the larger cars described earlier in the chapter were made ready for the Tour. It will be remembered that the old DB18 saloons which had been exhibited by Hoopers in 1949, had been disposed of when the first Tour was cancelled, as they were no longer current models. Some Daimler Consorts were also made obsolete after Princess Elizabeth cut short

Below: *The sturdy Conquest chassis went on to provide the basis for an infinate variety of models.*

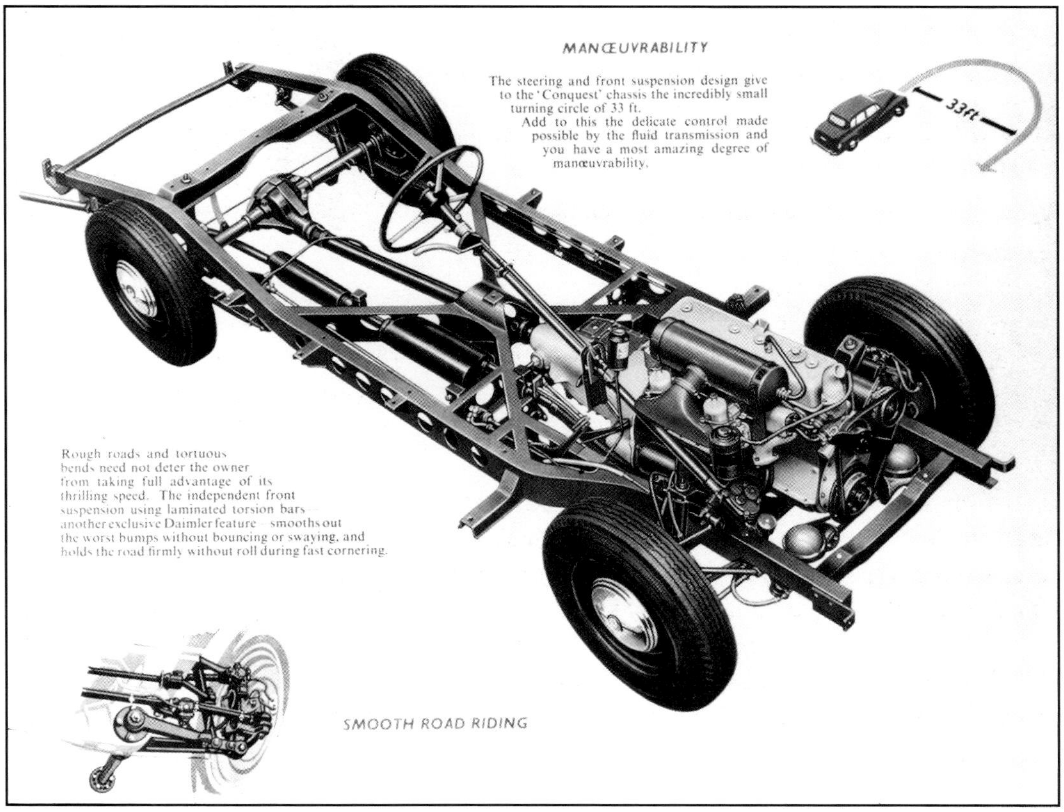

Above:
Daimler Conquest production in full swing.

Right:
A contemporary photograph of the Conquest Roadster; its engine was uprated first to 92bhp, and later to 100 bhp. With a few subtle modifications, the DJ254 transforms into an extremely capable machine.

Left:
The original Conquest Roadster (type DJ254), as shown at Earls Court in 1953. The Show car had a number of odd features such as ventilated rear wheel arch trims and no air intakes on the bonnet.

Right:
Of the two Lanchester Dauphins built, one was recently found in a derelict state in America.

her Tour to Australia in 1952. Some of these Consorts were in fact used, but when extra vehicles were required to augment the coachbuilt Royal fleet, Conquest saloons were usually employed. Apart from a few minor problems with the cars before the event started, everything went according to plan and the Daimlers once again proved to be very reliable.

Daimler diversification

By this time, Daimler had been involved in a couple of interesting projects, the first of which was a most unusual diversion for the Company - stationary engines. These "Unicorn" single cylinder diesel engines were rated at 10-12hp, and they were all exported abroad. The second project was even more unusual for Daimler, and came about through an order from the Ministry of Defence. Four 11.6litre diesel tractors were built; three with large wheels, and one with caterpillar tracks. Although the Company had produced paperwork for interested parties, it is thought these were the only examples ever sold.

The BSA Group managed to purchase the established Coventry firm of Carbodies in 1954. The extra capacity for specialist coachbuilding was extremely useful to Daimler, who were less than a couple of miles down the road. Carbodies had been founded by an ex-Daimler man - Bobby Jones, in 1926. By the late 1920s, the firm had already established itself as a fine coachbuilding concern, and by the 1930s, it was a major supplier to the trade of pressed steel panels. The company began producing its famous range of FX taxis in 1949, but it was the coachwork element of the business that interested BSA and, more importantly, Daimler. In the years that followed, "body by Carbodies" was to become an increasingly frequent sight on Daimler products.

Seven Conquests, suitably tuned to Conquest Century specification (of which, more later), had been entered on the 1954 Monte Carlo Rally. This severe test was enough to prove the chassis' worth, and five of the seven cars finished in the top 100, giving a fine showing on the Monaco Grand Prix circuit. Shortly after the rally, in the Spring of 1954, two new Conquest based models were announced; the Conquest Century - a high performance version of the standard saloon, and a smart Conquest drophead coupé. Both were in production by the middle of the year.

The change of policy was soon picked up in the *Motor* of June 1954: *"In the course of barely a year, the old-established name of Daimler has come to bear a new significance in which performance is accented as well as the traditional Daimler characteristics of quality and sound engineering."*

100 bhp Conquests

The Conquest Century was introduced in a bid to fill the gap in the range between the somewhat staid Conquest saloon and the racy two-seater Roadster. The Century engine was the same as the Roadster, and produced a healthy 100bhp at 4,400rpm. The propshaft, differential and braking system were modified to cope with the increase in power and performance, but otherwise the new chassis (allocated the designation DJ256, or DJ257 for left hand drive cars), was similar to the standard Conquest saloon (the DJ250) in almost every respect.

The external differences between the Conquest and the Century were few. The new model featured chrome plated surrounds to the windscreen and backlight in place of the usual black rubber, and vertical strips of glass were added to the rear of the front door windows in order to eliminate draughts. Small chrome flashes positioned on each side of the boot handle were also a notable feature on the Century.

Right:
The New Zealand Government commissioned Hooper to make a detachable perspex canopy for both cars that they already owned. This car (from the Royal stock, chassis no. 51717), was used in England to check the fitting for another car on the other side of the globe!

Left:
HM Queen Elizabeth II and the Duke of Edinburgh photographed while on the Tour of Australia and New Zealand. This picture was taken at Sydney Racecourse in 1954, and shows one of the superb Hooper Straight-Eight all weathers.

Some, although not all, had a Daimler badge fitted to the boot lid, and these were possibly the first Daimlers ever to carry the manufacturers name.

By far the most significant change inside the car, was the repositioning of the rear seat. After moving the seat further back, 4inches of extra legroom was obtained, and this modification was later applied to all Conquest saloons. The fascia board had to be re-arranged in order to make room for a tachometer, and the central panel was inclined slightly upwards. A telescopic steering column was fitted as standard, and an 18inches steering wheel replaced the 17inches one employed on the DJ250. Windscreen washers were included in the cars specification, with two-speed wipers being fitted to some examples. The front seats were also totally redesigned, resulting in more comfort and support. Praise was heaped on the Century from all corners of the world, and with a basic price of £1,172, it sold well at £1,661 9s 2d including purchase tax.

The Conquest drophead coupé (type DJ252/3) was a two-door four-seater with an electrically-operated hood, and had a public debut at the 1953 Paris Salon (the prototype Show Car was light green with a beige interior), despite not being available until the early part of the following year. This new and rather pretty model sported Century specification running gear, but surprisingly, no rev-counter was fitted. The hood could quite easily be left in the de ville position thanks to the straps provided, and with the hood fully erected, the cockpit was remarkably free from draughts. Unfortunately, as with most coupés of the period, rearward vision was restricted. At just £1,225 basic (about £1,736 with purchase tax) the DJ252 was very reasonably priced.

"Out of Pedigree Comes Pace"

The Century saloon was capable of covering a standing quarter-mile in under 21 seconds (depending on which road test one believes), and could reach 90mph. The Century went on to prove itself as an admirable competition car, and several examples were prepared by the Daimler works to be campaigned by Ken

Wharton, George Abecassis, and Reg Parnell. These particular cars were made to produce 116bhp at 5,800rpm.

At the 1954 International Touring Car Race held at Silverstone, Parnell had car no.11 (OHP 832), Wharton no.12 (OHP 831) and Abecassis car no.14 (OHP 828). Wharton led from the start after Stirling Moss stalled his Jaguar Mk.VII and the fluid-flywheel having been adjusted to aid a quick getaway. Sadly though, he was to crash into a spinning Lancia at Woodcote, thus ending his challenge. Jaguar won the event, but Parnell took 4th overall, and Abecassis came in 6th, gaining first and second respectively in their 2litre to 3litre Class. The Company coined the slogan *"Out of Pedigree Comes Pace"*, and as one magazine put it, *"The Daimlers, nicely turned out, had sprung a real surprise..."*

For 1955, the Company prepared three Regency Sportsman machines, including PRW 640 (3.5litres) and PRW 641 (4.5litres), to challenge the Jaguars - who were dominating the sport at the time - head on. The programme was cancelled, however, before the cars could be driven in anger on the track. It would doubtless have been an interesting year, as Westlake had been consulted on the engine modifications, but one has to wonder whether the Daimlers could possibly have competed on level terms with their Coventry cousins.

Above:
The frontal aspect of Silver Flash. Note the similarity between this cars grille, and that of the later Jaguar built 2 ½ litre V8s.

Right:
Daimler had built four tractors for the Ministry of Defence in 1953. This picture shows the only one with tracks.

Above:
The scene at Monte Carlo, 1954.

Right:
A 1954 Conquest Century saloon alongside a rather earlier example of the marque, a 1936 Straight-Eight limousine by Barker.

The 1954 London Show

Once again, Earls Court proved to be an important Show for the Daimler Company. It was decided to announce the 1955 Model Year range at the event, and no less than ten models were described for the coming season.

A fixed-head coupé version of the Conquest Roadster was exhibited for the first time by Daimler. At a basic price of £1,235 (about £1,750 with tax) a large number of buyers were expected, and hundreds of brochures were printed in readiness. The model was listed for several months, but only two or three examples were ever built.

Another, but fortunately more successful debutante, was the Regency Mk.II, or DF304/5. Although very similar to the old three litre Regency power unit, the Mk.IIs engine was enlarged to 3.5litres by increasing the bore from 76.2mm to 82.5mm. Maximum power went up from 90bhp to 107bhp despite a slight reduction in the compression ratio to 6.5:1. Virtually every other feature on this new 3,468cc engine was carried over from its three litre predecessor.

In spite of retaining the same wheelbase and track dimensions as the original Regency, this new version was a larger car - the sleeker bodyline adding about 5 inches to the length, further emphasised by a reduction of 2.25 inches in the cars overall height. The frontal aspect of the Regency Mk.II was made to look far more modern, and it differed considerably from the old model as can be seen when comparing photographs of the two. Inside the car, the interior was revised slightly to give more room, and picnic tables were added to the backs of the front seats.

Apart from extensions to cope with the extra length of the body, the chassis was virtually identical to that of the DF300. Girling hydro-mechanical brakes were employed, working on 12 inch drums all round. A larger fuel tank was fitted, but basically, the specification for the rest of the chassis was unchanged.

The basic price of the Regency Mk.II was put at £1,640 but purchase tax added nearly £700 to this. Although the car was undoubtedly designed with comfort as a priority, the performance of the new model was sluggish for a 3.5litre machine. In fact, there was little difference in performance between the Regency and the Conquest, despite the one-litre gap. It was hardly surprising therefore, that Daimler brought out a range of higher powered versions, and later, even a 4.5litre range.

The first of these was produced to form the basis for the ever popular Empress (by now in Mk.IIA form) and was designated DF306/7. The 3.5litre engine was made to develop 114bhp at 4,000rpm through the use of an aluminium cylinder head and a 1.1:1 increase in the compression ratio. The DF308/9 took the 3.5litre chassis one step further with a 130bhp engine, and one or two specialist bodies were mounted on it - the chassis also formed the basis for the Regency Sportsman.

In addition to the more powerful engine, the Sportsman differed from the standard saloon in several respects. The gearbox on the Sportsman (and Empress Mk.IIA) had a direct third gear, and an overdrive top ratio. The rear axle was also different, with a ratio of 4.89:1 compared to 4.3:1 as fitted to the standard model. Vacumn servo-assisted hydraulic brakes were fitted all-round.

The standard body for the Sportsman was that of a four-light saloon designed and constructed by Mulliners, but finished by the Daimler Company. The most notable features were situated at the rear end of the car; a large wrap-round rear window was employed,

Below:
Carbodies displayed this rather attractive fixed head coupé version of the Conquest Roadster at the 1954 Earls Court Motor Show. This example was perhaps one of two or three such cars. In the background to the left of the new model, a DJ252, also by Carbodies, is just visible.

Below:
The DJ252 Conquest drophead coupé, introduced in 1954.

Above:
The International Touring Car Race at Silverstone, held in May 1954.

and this blended in well with the overall body shape. The boot lid was set at a steep angle with the rear wings terminating in small fins, each housing a set of indicator and reversing lights, and a reflector. A continuous line ran the entire length of the car starting from the top of the front wings in a similar fashion to that seen on the Empress, and detachable covers over the back wheels added to the effect. In all, this design added about 2 inches to the length of the body.

Inside, the layout was much the same as the ordinary saloon, although on some examples the fascia was of the type which was later adopted on the so-called "Lady's Model". A 3.5litre Sportsman at the 1954 Show cost £2,650 including purchase tax, about £125 cheaper than a 4.5litre Regency saloon, and £450 cheaper than its big brother - the 4.5litre Sportsman.

The new six-cylinder 4.5litre chassis was much in evidence at the Show, with the larger Regency Sportsman (DF400/1), being shown on both the Daimler and the Mulliner stands. The 4.5litre version of the standard Regency Mk.II also carried the DF400/1 chassis designation, and this too was on display for the first time at Earls Court.

The DF400 (the DF401 was the left-hand drive model) used a 4,617cc version of the 3.5litre Regency Mk.II power unit. It was exactly the same in its general layout and design, except the bore was increased to 95.25mm, and the compression ratio was also slightly raised. After these modifications, the engine was made to produce a useful 127bhp at 3,600rpm. Gearbox ratios on the 4.5litre cars were almost the same as those used on the DF308/9 - third gear was direct, making an overdriven fourth gear.

For both the 4.5litre Regency saloon and the Sportsman, the standard 9ft. 6ins. wheelbase frame (as used on the 3.5litre cars) was employed. All the larger cars were fitted with a Girling servo-assisted hydraulic braking system, which was linked to a warning light mounted on the fascia to indicate the event of booster failure.

A new limousine - the Regina - was also exhibited. This was constructed on an extended 4.5litre chassis which had a wheelbase of some 10ft. 10ins., and used the same 127bhp engine as the saloons, coupled this time to a normal preselector-gearbox with no overdrive. The track measurement was increased on the limousine; the new dimensions being 5 feet at the front, and 5ft. 3ins. at the rear. There were several other mechanical modifications applied to the limousine; the Marles cam steering found on the saloons was replaced by a Burman re-circulating ball type unit; the brake drum diameter was increased by 1inch, and the tyre size was 7.50 x 16, as opposed to 6.50 x 16 and 7.00 x 16 found respectively on the 3.5litre and 4.5litre saloons. It should also be noted that a larger fuel tank was fitted, increasing the capacity to 20 gallons.

The Regina at Earls Court was painted in blue and black with a coachbuilt body by Hoopers. A fixed bench seat was fitted in the front compartment, and with the occasional seats, there was room for seven or eight passengers and a chauffeur. Standard features included a sophisticated, built in heater unit, an electric division, and automatic courtesy lights.

The Gold Car, the 1951 exhibit built on the DE36 chassis, was still attracting a lot of attention three years later at the new Daimler showrooms in

Left:
Dated September 1952, this picture shows an LJ202, or drophead coupé version of the Lanchester Leda. It is thought that the car was broken up before the public ever got a chance to see it.

Below:
Another debutant at the 1954 Motor Show was the attractive Regency Mk.II, available with either the 3.5 or 4.5 litre six cylinder engines.

Left:
This two tone Regency Sportsman was one of two that were prepared by the factory (with the help of Westlake) to race in the production car series. Unfortunately, they were never campaigned, but were later used by the experimental department as mobile testbeds for the Hobbs gearbox.

Below:
Elegance personified - the 1954 Motor Show car, Stardust.

Paris, situated on the Avenue Montaigne. However, the 1954 Show car (or Docker Daimler) was "Stardust" - another superb example of Hooper coachwork at its best.

Stardust was based on a shortened, and slightly narrower experimental DK400 chassis (no. 92700). A combination of silver and royal blue was used for the colour scheme, with hundreds of tiny stars in the lighter shade being employed to highlight the side panels below the waist. The well-respected motoring journal, *Motor*, described Stardust as *"... a masterpiece in the old sense of that word."*

Inside the car, the silver and blue theme was continued. The front compartment was trimmed in hand-woven silver silk brocatelle and blue leather piped with grey, and was designed to accommodate up to three people including the driver. In the rear, silver silk was again used extensively, and pale blue crocodile was used for much of the cabinet work and the door cappings. The cabinet itself was an elaborate creation produced in weight saving aluminium instead of the more usual wood. Drawers within the car contained everything that was needed for the occupants to eat, drink and be merry. The doors of the cupboards, situated on either side of the cabinet, folded upwards to form a continuous table almost the full width of the vehicle.

All four windows were double glazed and electrically operated, and a glass panel was fitted into the roof over the rear compartment which could be covered at will at the flick of a switch. Separate heating and ventilating units were provided for the front and rear compartments, and the large boot housed four matching crocodile suitcases and a vertically mounted spare wheel.

On the Lanchester side of the business, the Leda was replaced by the Lanchester Sprite. The Fourteen/ Leda run came to an end after about 2,100 cars had been produced - not nearly as many as had been hoped for. This was mainly due to the two litre engine, which was found to be slightly underpowered for such a heavy vehicle. In comparison, sales of the Daimler Conquest (a similar car but with a 2.5litre engine) were good, but instead of improving the Leda, the Company decided to opt for a completely new design.

The Sprite (code-named LM150) was of integral construction for the sake of lightness. The main floor formed a basis for the body from the rear seat pan to a massive bulkhead. Triangulated box-sections extended from this to take the engine and the independent front suspension. The transmission tunnel, front crossmember, wheel arches, and inner side panels, all combined to give the additional strength needed.

A new feature was that of engine access; the whole bonnet, along with the outer wings, could be swung upwards to make servicing easy. The engine itself was supposedly brand new, but was in fact just a four cylinder version of the proven Conquest unit, suitably modified by Harry Westlake. This work had the desired effect of giving the engine the same 60bhp found in the Leda, while losing 346cc and a substantial amount of weight. Further weight savings were made by building some of the body panels in light alloy.

Although fluid flywheel transmission was originally envisaged, it was decided at an early stage to better rival companies by offering the world's first medium-sized car with an automatic gearbox. Daimler had been toying with the idea ever since the war had ended - Wilson, Freebourne, and Sinclair units were all tried in the Experimental Department. It was the unique Hobbs system, however, which Dr Tait chose to develop.

The Hobbs gearbox employed a pair of friction clutches, a compound epicyclic gear and a whole series of friction brakes. Briefly, the gearbox was designed by Howard F. Hobbs for use in a personnel carrier just after the Second World War. The project was eventually abandoned due to a NATO amalgamation, but several companies backed Hobbs to carry on with the units development, and his son, John, was called in to help. The BSA Group was one of the companies who gave financial support to Hobbs, and Dr Tait was asked to give a detailed assessment of the invention. A number of points were raised and settled and, in 1955, the Group bought a 51% share in Hobbs Transmissions Ltd.

As time went by, a large order was landed to supply the German Borgward firm but unfortunately, after about 500 cars had been equipped with the Hobbs (or Mechamatic) gearbox, the company ran into financial troubles, and BSA promptly decided to sell its interests to the Westinghouse Brake & Signal Company. Further developments led to Ford being poised to use Hobbs transmission for their European light vehicles, but after a run of around 250 Cortina and Classic cars, they had a change of heart and that was the end of that. The gearbox, which had proved itself in everything from PSVs to delivery vans, and even several racing cars, deserved a better fate.

To return to the Lanchester Sprite, the Phase One Lanchester Sprite, made its debut at the 1954 Earls Court Motor Show, and was given a price tag of £760 plus purchase tax. Five chassis numbers were allocated for this model (70010 - 70014) but only three cars were ever built - one in all silver, the others being silver over maroon, and silver over dark blue. The lack of success can be gauged from the fact that not one of these was registered.

The Range for 1956

The Company's programme for 1956 was announced during the autumn of 1955, and there was a great deal of interest in the Daimler stand at Earls Court yet again. In the 3.5litre range, the Regency was renamed the One-O-Four, and given the new chassis designation of DF310/1. Incidently, the reason for the rather unusual name was because the prototype had reached 104mph in testing - this was no ordinary Regency!

The modified frame was made even stronger by employing a new I-beam cruciform member, and improved body mountings at the front of the car ensured a more rigid structure. Engine power had been increased from 107bhp to 137bhp at 4,400 rpm through the use of an aluminium head with improved gas flow, a new camshaft, and the latest SU carburettors. The rear axle ratio was also raised to 3.92:1 (or 3.64:1 if the optional high axle ratio was specified), and in addition,

Right:
After the introduction of the One-O-Four in October 1955, the Sportsman was renamed as the Four-Light saloon, even though it was basically unchanged from the previous year. This is, however, a strict Sportsman.

vacuum servo-assisted hydraulic brakes became standard equipment.

Burr walnut was now employed for the fascia, door cappings and picnic tables. The side armrests on the front doors were no longer fitted to give extra room, while those on the rear doors were modified slightly to allow easier access. Tread plates finished the four doorsteps, and the courtesy lights were made to work when any door was opened.

The spare wheel tray was also of a new design, making it near impossible to see from the rear of the car, and the boot gained an extra 2 inches in length. New flatter hub caps were employed, and a D medallion with traditional Daimler fluting now graced the rear bumper. All these modifications combined to transform the Regency Mk.II, making the One-O-Four an outstanding machine. A basic price of £1,885 was given to the new model, or £2,670 with purchase tax.

The Regency Sportsman continued to be produced, but was now known simply as the "Four-Light" saloon, built on the standard DF310/1 chassis. In its final 3.5litre form, the Empress (in Mk.III guise) was always built on the old DF308 chassis, fourteen of which were made. An interesting variation on the DF310, however, was the so-called "Lady's Model" - a One-O-Four fitted with a wealth of extras designed to please the lady motorist.

The Lady's Model exhibited at Earls Court (chassis no. 96851) was finished in silver and black, and came away with a Silver Medal in the coachwork competition. The standard equipment included a vanity case full of cosmetics, a rolled gold pen, cigarette holder, a notepad, torch, sunglasses, collapsible umbrella, travel rug, a map, shooting stick, picnic box, and his and hers suitcases. Electric windows were standard, and the dashboard was redesigned with all controls and warning lights identified by name plaques. In their road test of the Ladies Model, the *Autocar* noted, *"... he is a fortunate husband whose wife owns such a vehicle."*

The 4.5litre saloons were given a little more power for the 1956 trading season, and the chassis was given the new designation of DF402/3. Several minor modifications boosted the engine output to 167bhp at 3,000 rpm but unfortunately, the torque proved to be too much for some of the overdrive gearboxes, and several 4.5litre chassis were either withheld or recalled; it is possible that the chassis frames were later used for 3.5litre saloons. A superb royal blue and silver 4.5litre Empress was displayed on the Hooper stand - one of the most desirable of all post-war Daimlers.

A similar increase in power was given to the 4.5litre limousine engine, although 3,800rpm were needed to achieve the same 167bhp output. A slightly revised limousine was built, known simply as the DK400 (the name Regina had been dropped after just one coachbuilt car had been produced). The original Regina was retained by Daimler for a while, and was later used by the Royal Family during their Tour of Sweden in 1956.

Daimler decided to ask Carbodies to produce a standard body for the 4.5litre DK400 chassis. The overall length of the car remained the same at 18ft. 1ins., and interior changes between the new model and the Regina were few. When the DK400 made its debut at Earls Court, it was priced at £2,793 basic (or about £3,958 with tax), compared to the Regina's basic price of £4,385.

Hoopers, however, came forward with an alternative design to the standard, somewhat ungainly, DK400. The Hooper design resulted in a superb limousine capable of seating up to eight people, constructed with the extensive use of light alloys. A laminated windscreen was fitted as standard, as was a radio and separate heater for the rear passengers. In most cases, two forward-facing occasional seats were provided, which folded flush against the division when not in use. The price for this splendid machine was set at £4,385 plus purchase tax - the same as the old Regina.

In all, just seven of these cars (Hooper design no. 8434) were built. The first one was the Show Car, and carried body 10148 on chassis no. 92706. The Lord

Left:
The Beutler bodied 3.5 litre Daimler convertible in its native part of Switzerland.

Top of facing page:
The 4.5 litre Daimler Regina limousine by Hooper.

Right:
A Phase I Lanchester Sprite on test, similar to the contemporary Singer 1500 models in profile.

Left:
George Fabel and Cyril Corbishley on the 1955 Monte Carlo Rally.

Right:
The shape of things to come - Hooper body on the new 4.5 litre DK400 chassis. One similar to this example (body no. 10152 on chassis no. 92715) was supplied to Princess Marina.

Mayor of Cardiff had the next one, and the registration "KG 1", which had been seen on several of that City's Mayoral Daimlers before, was transferred to this latest acquisition. Another one went to HH the Sultan of Johore (body no. 10149 on chassis no. 92708), while Princess Marina took delivery of body 10152, chassis no. 92715. The remaining three Hooper cars (bodies 10150, 10151 and 10153), all went to private customers, the last being delivered in October 1958.

A car of similar design had been produced for Her Majesty the Queen Mother - this was to be the last Daimler ordered by the Royal Family for well over a decade. Since this vehicle (chassis no. 92702 with body no. 10135) was not a State car, it was painted all black, and trimmed in walnut and West of England cloth throughout; no occasional seats were fitted. As with the King and Queen's earlier personal cars, the radiator carried a lion mascot.

The Golden Zebra Car

The Golden Zebra car was to be the last of the almost legendary Docker Daimlers. Based on the DK400 chassis (no. 92705), the Hooper body (no. 10146) was finished all over in plain ivory-white, with the parts which normally would have been chromed, plated in real gold. Many of the interior fitments were also gold plated, and ivory and zebra skins were used extensively throughout. On being asked why Lady Docker had chosen Zebra skins, she is said to have replied, *"It's the best skin, because mink is too hot to sit on."*

The front seat was of three separate panels pushed together to form a bench, and in the rear, there

Right:
The last Docker Daimler - the 1955 Golden Zebra car.

Below:
The Golden Zebra car as it is today. At the time of writing, the vehicle was in England, and owned by John Wentworth who runs his own Daimler spares business in the Cotswolds.

Below right:
A special Empress style body on the Mk. II Century chassis.

were two folding armchairs. All the seat faces were covered in real zebra skin, while the backs and sides were trimmed in the finest quality ivory coloured leather. The headlining was also ivory coloured with tiny gold spots and, as with the earlier Docker cars, a perspex panel and shutter was fitted in the roof. The car was fitted with every imaginable extra, and the finishing touch was a small gold zebra mounted on top of the famous fluted radiator. Reputed to have cost around £12,000 to produce, Lady Docker had set out to *"show the world that we* [Daimler] *had craftsmen who could build the finest products in the world."*

The last Lanchester

One of the most interesting production models on display in 1955, was undoubtedly the Phase II Lanchester Sprite. It will be remembered that the original Sprite had been introduced at the 1954 Show, but had little success. This new version however, looked far more promising, and was expected by the Daimler Board to be a best-seller.

The Phase I Sprite had a similar shape to the contemporary Singer, but the Phase II employed many of the panels found on the Daimler Conquest. Some assert this was for cheapness, and others believe it was because of the delay involved in making new parts, but in any case, the result was a much prettier machine.

The Phase II resembled the Conquest in other ways, too. The coil springs of the earlier Sprite gave way to long torsion bars anchored to a transverse I beam halfway down the body. From this beam, two legs of an embryonic chassis extended rearwards to take the back axle, and two deep box-sections ran forward off it to

Above: *The New Drophead Coupé by Carbodies - first seen at the London Motor Show in 1955, and winner of a Gold Medal for coachwork on its debut.*

Below:
Commercial vehicles played a large part in the fortunes of the Company. Note the new style grille, which in fact makes the vehicles look more dated than their predecessors.

give additional strength. This time the bonnet was of the conventional type, as the large one-piece bonnet and wing structure had proved to be a better idea in theory than in practice. The battery was moved into the engine compartment, although an indentation could still be found in the rear seat pan, where it had been housed on the Phase I. The rear wings, windscreen and rear window came directly from the Conquest.

Three Phase II prototypes were built; RKV 146 had been used for the handling and fuel consumption tests, and was finished in grey over black with a red trim. The two other cars sported cream and deep red coachwork respectively, and were both fitted out with a red trim. The customary 1,000 miles on the MIRA test track had actually been done in stages, as on several runs, parts of the car had become dislodged. Torsional rigidity also caused a few problems and stiffeners were required to obtain a satisfactory result. Gearbox judder was solved by a few phone calls to the BSA motorcycle engineers, who suggested making the clutch lining in a different material. Fortunately, all the teething troubles were soon put right.

Despite increases in both the wheelbase and front track dimensions to 8ft 4ins and 4ft 4ins respectively, better ground clearance and slightly larger tubeless tyres, the later Sprites still managed a tighter turning circle than that of their predecessor.

Cost cutting measures included a reduction in unwanted chrome and the loss of the central chassis lubricator. Aluminium gave way to steel for the door panels, and much of the trim was made of cheaper materials than Dr Lanchester would have liked; the front bench seats, for example, were finished in plastic instead of leather, reflecting modern trends. Optional colour schemes became a little more adventurous too, and included such combinations as pastel blue over mid-blue, and stone beige over scarlet.

At the Show, two Sprites were put on display; one in silver over black, and the other one in silver over maroon. The first deliveries were expected in mid-1956, and a price was quoted of £1,230 including tax. Sales brochures and owners handbooks had been prepared, and in the meantime, production planning and development continued. An SU carburettor was scheduled to replace the Zenith one after the first 500 cars, and after 1,000 units, several steering modifications were planned. Future models were to have included higher powered versions, and even left and right hand drive convertibles were not out of the question, as the chassis references LM152 and LM153 had both been reserved for at least one of these projects.

The Mark II Conquest

The 1955 Show created an opportunity to introduce changes in the Conquest range of vehicles, and so the Conquest and Conquest Century saloons made their respective debuts in Mark II form. Although there were no significant mechanical modifications, there were several minor changes elsewhere. The rear suspension was revised, and a new Smiths heating and ventilation system was employed. Burr walnut was used on the fascia and door cappings, and the instruments themselves were moved in the course of production; black faced dials gradually superseded the old white ones. Separate fog and spotlights were mounted just above the bumper, and the holes that this arrangement left either side of the grille were converted into air intakes. As Motor put it: *"The Century is a car which appeals strongly to the connoisseur on the grounds of engineering, refined behaviour, and high quality of detail and finish."*

By far the most important change in the Conquest range was the introduction of a new model - the New Drophead Coupé. This was a cross between the DJ252 and the DJ254 - the two door four-seater coupé had been discontinued after about 235 had been built, and the Roadster name had disappeared completely. After this, the DJ254 continued to be built, but in a reviewed form which was basically a much improved three seater version of the Roadster. The seats were arranged in such a way that the front passenger seat could be hinged forward to allow access to the single rear seat. This back seat was positioned sideways, but it did have the advantage of being full sized, and could also be folded or even removed to allow for more luggage space.

The New Drophead Coupé continued to use the standard Century chassis and engine, although an alternative higher rear axle ratio of 3.73:1 was made available as an option. The body was produced by Carbodies, and now featured wind up windows and a slightly revised waistline area. These modifications impressed the Show's judges, and they awarded the new car with a Gold Medal for the best coachwork in its Class.

It is interesting to note that towards the end of their run, a few New Drophead Coupés were sent to be bodied by Airline Airstream Ltd of Northampton, but instead of the usual aluminium, the material employed by this company was glassfibre. It is thought that these cars were built to provide Daimler with some data regarding GRP bodies, as it would be useful knowledge for a future project known as the SP250.

The 1955 Earls Court Show had once again proved to be a great success for the Company, but although everything looked to be running fairly smoothly from the outside, the Daimler Board were becoming increasingly concerned with the behaviour of the Docker's.

CHAPTER TEN

THE END OF AN ERA

The mid and late 1950s were to signify the end of an era for Daimler in more ways than one. Sir Bernard Docker became the object of corporate criticism, and was eventually ousted from the Company, apparently because of his complete lack of communication and outrageously extravagant lifestyle. This latter accusation was wholly justified, as hardly a week would go by without Sir Bernard, Lady Docker, their parties or their £100,000 yacht, Shemara, being in the headlines, and most of this was being charged to his expenses on the grounds that Lady Docker's glamorous image was good for the Company.

The downward slide had begun in 1953, when the Midland Bank forced him to resign his senior Directorship - his profile was far too high for the Bank's liking, and his antics had even come to light in the House of Commons. A Court case brought against Sir Bernard involving an alleged currency offence shortly after, did nothing to help his situation, and he resigned from the Board of the Guardian Assurance Co., giving no reason for his decision.

The Paris Showroom disaster (where Lady Docker spent nearly £8,000 on outfits for the occasion), and the fact that Sir Bernard had used Company funds (over £2,000) to fly Stardust and the Golden Zebra car to Monaco, so that Lady Docker could attend Prince Rainier's wedding celebrations, infuriated the Board. *"It is in the opinion of your Board; that this type of publicity did not produce any reasonable return in commercial advantage or prestige value to the BSA Group"* ran their statement.

Facing page: *Early 'A' spec. Daimler SP250s on the production line.*

FIRE
Within

Matters really came to a head when Docker tried to make his wife's brother-in-law, R.E. Smith, a BSA Director. Although this proposal was rejected by five votes to four, Sir Bernard used his casting vote to declare it passed. This proved to be his final victory, for on 30th May 1956, his Chairmanship was overthrown in his own home. Lady Docker saw it as a big plot from the start. She said that Sangster (the man that took Docker's place), James Leek (BSA's Managing Director and a Director of Daimler since 1948), and Charles Rowe (the Director & Accountant), had dreamt up a scheme which would displace her husband.

In the words of Lady Docker, *"Later, of course, I heard the full story. How six Directors had voted with the resolution to sack him, against three who had supported him. The two that let him down were men he had not imagined were against him. One was his own cousin, Noel Docker, and the other, Sir Frank Smith, a scientist, who had been his father's closest friend."*

Following confirmation of his fate in August, Sir Bernard incredibly bought some time on Independent Television to appeal against the decision directly to the BSA shareholders, and Lady Docker did a sterling job in the Press, fighting on principles. Whether the public were convinced that Sir Bernard *"makes very little from BSA"* on nearly £45,000 a year or not, doesn't really matter, for the Docker's had forced an EGM.

In Lady Docker's words, *"One of the strangest features of the affair was the intervention of the Prudential Assurance. The Prudential were the largest shareholders in the Company, with 218,000 units - Bernard and I held 120,000 between us.*

"Normally, it is considered highly unethical for other companies to interfere with the internal affairs of an organisation in which they hold an investment. For this reason, the Prudential's man at the meeting, their investments manager, Leslie Brown, came in for a great deal of criticism from certain quarters in the City.

"Bernard and I were particularly concerned that the Prudential should have seen fit to take sides in the dispute. Why were they supporting the new Board, who had so few shares? Maybe the Prudential people had been influenced by the vitriolic campaign launched against us. Whatever their reasons, the massive vote they controlled weighed overwhelming against us.

"There were prolonged shouts of disapproval at the meeting, when Jack Sangster was forced to admit that he only held 2,000 BSA shares. But he was occupying the Chair, and there was no way to remove him."

Despite another attempt at re-election at the

Above:
A Carbodies limousine on the 4.5 litre DK400 chassis.

Above: *Jack Sangster.*

Above: *Edward Turner.*

AGM, held in December 1956, nothing came of Docker's plea - they simply resigned from the BSA concerns in which they were involved, moved over to Rolls-Royces, and continued their high-profile lifestyle almost as if nothing had happened, but it was undoubtedly a sad affair. *"Pushed up the business ladder too soon"*, and having wasted the family fortune, Sir Bernard died in May 1978 after a long illness, while Lady Docker was still sprightly until she passed away in December 1983.

John Young "Jack" Sangster, had joined the BSA Board in 1951. Born in Kings Norton in 1896, Sangster was a shrewd businessman who had spent most of his life in the motor trade, building his own two cylinder air cooled machine just after the First World War. Rover bought the manufacturing rights, and Sangster was given the job of organising production. After making a name for himself with Rover, Sangster went on to join his father and introduced cars such as the Ariel Nine and Ten. By the 1930s, Components Ltd were in serious financial trouble and the Receiver was called in. Sangster bought all Components Ltd's plant and a part of the factory, and started trading as Ariel Motors.

Shortly afterwards, in 1936, Sangster bought the Triumph motorcycle business, and eventually both Ariel and Triumph were sold to the massive BSA Group. Sangster was the man who brought in Edward Turner, previously of Triumph, as the Managing Director of BSA's newly formed automotive division, and both men proved to be of equal importance to the Daimler Company in later years.

Turner, born in January 1901, had always had an interest in motorcycles. After some notable work at Ariel, Jack Sangster put Turner in charge of his new Triumph motorcycle business - his designs turned the company around, and so when Triumph was sold to BSA in 1951, Sangster managed not only to get himself on the BSA Board, but Turner his new post as well.

Before moving on, it must be pointed out that Daimler's financial failings were not due entirely to Sir Bernard and Lady Docker. For instance, all BSA Directors had Daimler limousines after the war, and none of them complained at the time. Purchase Tax changes made the cars too expensive for the middle class or retired buyer - not Daimler's fault - but there

Above:
The elegant lines of the 1956 Continental saloon by Hooper.

Right:
The Phase II Sprite spelled the end of the Lanchester marque. This example shown here is the only survivor, and is pictured with members of the Hobbs family and John Ridley (in the car) who is the current owner of the car.

Left:
Reminiscent of the Golden Zebra car, this special Hooper bodied DK400 was built for the Emir of Bahawapur.

Foot of page:
A 1956 Royal stock DK400 limousine. Original intended for the use of the Queen Mother and Princess Margaret, the car was never delivered to the Royal family, a Rolls-Royce taking its place.

was no effort made to remedy the situation by making the small cars cheaper, or the problem in the market hadn't been seen at all. Surely, Docker cannot be held solely responsible for this.

In addition, as Micheal Sedgwick pointed out: *"Daimler played the role of David against three Goliaths, challenging Rover with their Conquests, Jaguar with their sports models, and Rolls-Royce with their eight-cylinder DE36."* Without the necessary financial backing from BSA, the cars could sadly not be developed to keep pace with the competition. The inevitable happened, and *"they lost all three battles."* Perhaps the biggest mistake during the Docker period was neglecting the Daimler bus side of the business, for in this field the Company had a significant lead over all their rivals, but gradually let Bristol and Leyland overhaul them.

All change

One of the first things that the new regime introduced to the Daimler range, was the Borg Warner automatic gearbox. The Daimler transmission had been in use for over 25 years, and was still a perfectly good unit in service. However, the automatic gearbox was now becoming fashionable in this country, and the management decided that it was time for a change. Initially, the Borg Warner automatic gearbox was offered as an optional form of transmission, but it was not long before it became a standard fitment. The new Board preferred to use the proven Borg Warner than to keep pumping money into the Hobbs transmission, even though it was nearly ready for production. At about the same time, the coachwork of the DK400 was completely revised. The DK400 retained its four door six-light specification, but the lines were now of a more modern appearance.

With the Docker era coming to such an abrupt end, it was unfortunate that one of Docker's most promising projects, the Lanchester Sprite, was not developed further. Several customers had already been told that it would not be long before they could take delivery of their cars, but the new management decided that it would cost far too much to put the Sprite into

production. Many people who were involved with the project felt this was just an excuse to drop the model, as virtually all the teething troubles had been eradicated, and the Sprite had all the attributes of a winner. However, by the time that the 1956 Earls Court Motor Show had arrived, the Sprite had totally disappeared from the market, and the Lanchester name with it. When the Hobbs family departed from BSA, they took their drawings with them, as well as most of the remaining Hobbs gearboxes, so the last few Sprites left at the factory were fitted with preselector transmissions and put to use as works hacks.

In all, a total of only thirteen (chassis numbers 70100 - 70112) Phase II Sprites were built; three prototypes, and ten production models - a sad end for the Lanchester marque. Today, only one Sprite is known to exist. Registered THP 476, it is the last true Lanchester. The whole Sprite episode is said to have cost BSA £500,000.

Earls Court, but no Docker Daimler

The two door Hooper bodied Continental first appeared at Earls Court in 1956 and made a big impression. Originally, the 4.5litre type DF402 chassis was to have been used for the car, but it was decided that this would be an ideal opportunity to demonstrate the newer DF314 chassis - a One-O-Four with automatic transmission.

When the idea for the Continental was first conceived, it was planned as a direct competitor to the superb but more expensive Bentley Continental. With this in mind, many of the features found on the Bentley were incorporated into the Hooper body - in particular, measures which excluded draughts and dust, and to ensure insulation against noise and fumes.

Inside the car, two adjustable bucket seats with hinged backs were employed for the front occupants, while at the rear, there was luxurious seating for two, or occasionally three people. When not being used for passengers, the rear seats could be folded out of the way to allow for more luggage space. Eucalyptus burr was used extensively within the car, and although no tables were provided, creature comforts were still very high on the list of priorities.

The Continental was 17ft 3ins. long and because of this, the chassis frame had to be lengthened slightly. The body was built to be as light as possible without loss of strength, and the total weight of the car was just over 35cwt - considerably less than that of a standard One-O-Four saloon. It was hoped that the vehicle would be put into limited production, but unfortunately the Continental project was never developed (probably because of the Suez Crisis) and only one was ever built. However, the car was not scrapped and has recently undergone an extensive restoration in the hands of a true Daimler enthusiast.

Below: *This DK limousine (chassis 96015) was first painted black with the body sides and bonnet finished in a dark beige colour. Its debut came at the 1957 Motor Show were it was exhibited on the Daimler stand. The car is now part of the Jaguar-Daimler Heritage Trust collection.*

More DK400s

During the April of 1957, the Emir of Bahawalpur took delivery of his specially-bodied DK400 from Hoopers. Finished in cream, the car was built to resemble the Golden Zebra car which he had admired

Right:
Possibly the only drophead coupé ever built on the DK400 chassis, this splendid machine was built for the King of Afganistan in 1957.

at the 1955 Motor Show. Chassis 92716 was used and the body (no. 10235) had several interesting features. By far the most striking of these was the roof which was actually a large purpose-made canopy of transparent moulded perspex.

The body itself was constructed using aluminium and several alloy castings, some of which were specially made for the car. Extra sound deadening and insulation materials were added to the floor and doors, and a barrier of aluminium and asbestos was built up above the exhaust system. The interior of the Emir's car was finished in cream leather for the seating, and birds eye maple was used for all the cabinet work, fascia and door cappings.

The next body built by Hoopers (no. 10236) was also a very special one. This was to be a superb all-weather car for the King of Afganistan, and was probably the only drophead coupé ever built on the DK series chassis. The chassis number was 92724.

After the first two Hooper bodied DK400 cars, Stardust and the Regina, hooded headlamps became a distinguishing feature of all further DK400 bodies built by the famous coachbuilding concern. In fact, only one other non-DK series Daimler was to share this feature - the 3.5litre Continental show car.

At the 1957 Motor Show, the DK400 was shown with two alternative seating arrangements. The original three-abreast occasional seats were still available (this model became known as the DK400A), but for those customers who demanded the maximum amount of comfort, a new design was put on the market. This version (the DK400B) had occasional seating for two via a pair of greatly improved folding seats. Both seats were 2inches wider than those previously fitted, and designed to disappear completely into the special division. The division in the B specification cars had a concealed winding mechanism, it being of the drop type, compared to the sliding glass used in the original limousine. At the time of the Show, the DK400 sold for £4,195 including tax, and the DK400B was priced at £4,315.

The End of an Era

Before 1957 came to an end, the Docker Daimlers left at the factory were broken up. These included Stardust, which was *"rendered anonymous"* then re-registered and sold - Simmons of London W1 had the car up for sale in April 1958 for just £3,750. The Golden Zebra car was also toned down before it was put on the market. All the gold fittings were removed, including the small zebra radiator mascot, and the coachwork was painted black. The Gold Car went to the U.S.A. - it was a sad end for some of the most expensive cars ever built.

The Conquest Century saloon was marketed with automatic transmission only in the last months of 1957. The use of the fluid flywheel, on the smaller Daimlers at least, was a thing of the past. In January 1958, the last of the highly successful Daimler Conquest Centurys was produced, although it is known that one or two left the factory after this, built up from spares. In its fairly short life-span, over 4,000 Conquest Centurys had been produced.

The last of the Empress line was also produced in 1958. This famous Hooper body style had first appeared some ten years earlier on the DB18 Special Sports chassis, and proved so popular that it was adopted in various forms on 2.5litre, 3litre, 3.5litre and 4.5litre Daimlers, gathering many prizes in coachwork competitions along the way.

The modern Daimler

In July 1958, the Majestic was introduced to the Daimler range as a direct development of the proven 3.5litre One-O-Four. The Majestic was given the type no. DF316 for right-hand drive cars, and DF317 for vehicles with left hand drive. The car had a completely restyled flush sided body which, helped by a lower radiator line, not only looked more modern than its predecessor, but also offered more room for the occupants. The whole bodyshell, built by Carbodies, was increased in width from the scuttle rearwards, so that the line of the front wings was made to end in the front doors. From this point, the body sides were continued at full width to the rear of the car.

Inside the Majestic, it was typically Daimler. The upholstery was deep and comfortable, and there were central armrests in both the front and rear bench seats, as well as further armrests fitted to each of the four doors. The amount of passenger space available was substantially greater than it was in the One-O-Four, and additional legroom was provided in the rear by a recess in the back of the front seat.

Mechanically, both the engine and braking system received attention, and only automatic transmission was offered. The new engine was a six cylinder unit of 3,794cc - this was not achieved simply by boring out the One-O-Four block, but by making a completely new casting. With a bore and stroke of 86.36 x 107.95mm, the block itself had been sufficiently enlarged to allow reboring and, if necessary, a second rebore with the fitting of liners. Unlike previous models dry liners were not used.

Again, an aluminium head was employed, with a much improved gas-flow. Raising the compression ratio to 7.5:1 resulted in a 10bhp increase in maximum power; at 4,400 rpm, 147bhp was now obtained, and the engine's torque characteristics were also much better suited to the car's transmission. The automatic transmission featured a torque converter which locked out in top gear, and had a provision for manually holding the intermediate ratios.

The cars braking system was very interesting. The Company had been experimenting with disc brakes since about 1952, and now Daimler was at last confident enough to offer this form of braking as a standard feature. In fact, the Majestic was almost certainly the first British saloon car to have disc brakes fitted as standard, although Jaguar were already fitting them to their XK series sports cars, and offered them as an option on their saloon car range. The brakes were of Dunlop design, operating on 12inch diameter discs at the front and rear of the car. A vacuum servo unit was employed, and the handbrake worked on a separate pair of callipers at the rear.

Daimler decided against using automatic chassis lubrication on the Majestic. Automatic chassis lubrication had been a feature on most of the Company's cars since before the Second World War, but it was decided to increase the number of grease nipples instead. The reasoning behind this reversal of tradition was sound, because if any part of the automatic system failed, and remained undetected, sooner or later further difficulties would be encountered - grease nipples could be easily maintained each time the car was serviced.

The *Motor* magazine summed it up as follows: *"Altogether, this Daimler Majestic is a car which fills an admittedly limited, but nevertheless very real, market which has been neglected in recent years - a market for a car which is frankly large and luxurious, but which matches these requirements with speed, acceleration, and handling qualities that go far to minimise the effects of sheer size in the congested conditions of today. With this combination of qualities, its basic price of just over £1,662 must be adjudged moderate."*

Autocar followed on by saying: *"It has a quite excellent performance for its size, with first class brakes and good roadholding - in a word, it is in all these respects the best Daimler model which we have road-tested."*

Shortly before they closed down in 1959, Hoopers carried out several special orders for customers on Majestics, although most of this work involved nothing more than the fitting of divisions, and refinishing cars with new paint schemes - no specialist coachwork was ever fitted to the Majestic chassis.

The SP250

One of the most interesting of all recent Daimlers, was the SP250 (or "Dart" as it is more commonly known). Introduced in April 1959, the Dart was a complete departure from the Company's standard practice. To understand why the SP250 was introduced at all, we must first go back in time, and look at what was happening in the Company's Experimental Department.

For two or three years, Daimler had been looking into the possibility of cutting costs by, if you like, "Daimlerising" a mass-produced car, and employing a brand-new small Daimler engine. The Panhard Dyna 54 was given a great deal of consideration, but unfortunately, the Government of the day were totally against the importation of ready made panels. Despite the enormous size of the BSA empire, it did not have the capacity needed to manufacture its own aluminium body panels, so the idea was dropped.

As an aside, one of the very last Jowetts found its way onto the Daimler testbed. As this marque was already extinct, it seems likely that the engineers had

Below:
An early advertising proof for the new Daimler Majestic.

Above: *A superb Empress III coachbuilt body mounted on a 1956 DF308 chassis. Many enthusiasts have mourned the passing of the Empress line.*

obtained one to see if they could crib any ideas from it, and were not intending to reproduce it. It also seems that the Company was at one time considering a front wheel drive three cylinder two stroke machine, which would have changed the course of Daimler history beyond recognition.

However, after the Panhard idea failed to materialise, Edward Turner entered into private negotiations with Vauxhall, with a view to producing a Daimler built Vauxhall Cresta-based saloon. A chassis designation of DN250/1 was assigned to the new project and in the meantime, Turner, who was the BSA Group's engine designer, had the new small power unit poised to go into production.

Development work on the new engine had started late in 1956, and by the summer of the following year, the first unit was built and running. The engine was a 90 degree V8 with a capacity of 2,548cc. It employed a cast iron block, aluminium alloy heads with hemispherical combustion chambers, and aluminium alloy pistons with steel connecting rods. A nice feature was the high mounted single camshaft/pushrod arrangement which managed to keep the engine a very compact unit. The very first V8 was fitted with a twin-choke Solex carburettor, although it was later tried with eight Amal motorcycle carburettors. The Amal carburettors resulted in an overall increase of 60bhp, but after further experimentation it was eventually decided that two SU HD6 carburettors should be used when production finally got under way, giving a very respectable 140bhp.

The Daimler engineering team was headed by the vastly experienced Cyril Simpson and the initial road testing of the engine took place in a Conquest Century. The new V8 proved itself over more than 12,000 miles, and the design appeared to be a good one from the start. However, the Company was left with a problem. One prototype DN250 had been built, but the contact in Vauxhall was still holding proceedings up. They now had a superb 2.5litre engine, but nothing to put it in.

All that Daimler had on the market, was the large 3.8litre Majestic and the somewhat dated DK400 limousine, both of which were built in relatively small numbers, especially the latter. The Majestic, good though it was, could certainly not be relied upon to carry Daimler through these hard times. Thus, a new car was needed which had to be right first time. Rather surprisingly, the management decided to build not a saloon, but a sports car; presumably they felt that the deal regarding panels might still come to something. In any case, the SP250 project was born.

With cash and time in increasingly short supply, as many shortcuts as possible had to be taken. The pressures of the time dictated that a chassis design would have to be borrowed from somewhere. So to speed things along, a white Triumph TR3 was purchased and stripped down behind the closed doors of the Radford works, with all those involved being sworn to secrecy; Bob Guillianotti took the car to pieces, and Phil Lomas drew each of the TR components in turn.

The chassis specification included coil and wishbone front suspension, an underslung rear axle on semi-elliptic springs, and cam and peg steering. The braking system, however, was designed especially for the new car, and employed the use of Girling discs on all four wheels.

The original SP styling was by Jack Wickes, *"Edward Turner's pencil"*, and then left in the capable hands of Percy McNally. He was the General Manager of the Carbodies plant at the time, and had previously been responsible for the styling of the Majestic and the Phase II Lanchester Sprite, as well as the infamous Austin FX4 taxi. The resulting lines were quite unlike anything Daimler had come up with before, but overall, the effect was rather pleasing.

Due to the excessive time and expense involved in making steel bodies, the BSA Board decided at quite an early stage that the shell should be made of glassfibre. This would keep the tooling costs down, and make use of the existing facilities owned by the Company, which made panels for the Daimler buses - this course of action also enabled production to get under way far quicker. The use of glassfibre may have only been intended as a temporary measure though, as a statement was later released saying that should demand for the vehicle increase, the body would almost certainly be built in steel.

At about this time, a survey of projected SP250 sales was prepared. It was hoped that something like 7,500 cars would be sold during the first three years of production, with about 1,500 being sold in the first twelve months. If met, these figures would have made the car one of the most successful Daimler models ever made. It was vital, however, that the Daimler Dart was a success. Even if the sales targets were met, the Company still wouldn't see a profit until the end of the third year, and a heavy loss of about £500,000 would be experienced during the second year.

The original testbed prototype (illustrated here, registered WDU 654) was driven extensively around the Welsh countryside. The prototype was fitted with an Austin-Healey gearbox, as opposed to the four-speed Triumph TR based unit that production models were equipped with. The prototype's body was merely a mock up of what was to come, and bore only a passing resemblance to the finished product. According to John Box, one of the first Daimler people to actually take the machine out for a test run, the car was a *"real animal"* to drive, and he could vividly remember several incidents with the car that he would much rather forget! The driver's report on the car recommended several minor changes to the steering and suspension, as well as a change of tyres from the Dunlop RS3s used on the first test car.

After several months of testing, the time at last arrived for the official launch of the new Daimler. The 1959 International Automobile Exhibition, held in New York during April, was the chosen venue. It is not clear why America was chosen for the launch, as to this day, the Daimler marque is still virtually unknown in the U.S.A. However, export hopes for the Dart were high, and it is possible that the management hoped to win extra sales from the goodwill shown by launching such an important model in America, along with the added appeal of a V8 engine.

Certainly, the Company went to a great deal of trouble to gain good public relations with the American's. Earle C. Antony, a local car dealer, actually managed to keep one of the runways at Palm Springs Airport clear for over an hour, so that he could demonstrate the speed of the car to potential customers. The same car had just completed a 15 day tour across a large part of the country, just to prove that it was as reliable as it was quick. No less than 6,500 trouble-free miles had been clocked up by the time the tour came to an end.

Series production of the SP250 began in September 1959. By now, Daimler had been banned from using the name "Dart" as Chrysler had already registered the name, and objected to its use by other manufacturers. There were a few minor changes before production; the grille was made slightly shallower, a lid was now fitted to the glove compartment, and tailor-made rear lamp units replaced the Alvis ones originally used. Including purchase tax, the launch price of the SP250 was put at £1,395 - slightly higher than was hoped, but nonetheless, still a good buy. At this stage,

Left:
The Daimler Majestic, introduced in July 1958.

Left:
Jack Wickes drawing of the proposed DN250 saloon dated November 1958.

Right:
A styling mock up of the elusive DN250 saloon.

optional extras were to include such items as front and rear bumpers, wire wheels, whitewall tyres, an adjustable steering column, tonneau cover, a detachable hard top, a leather covered steering wheel and Borg Warner automatic transmission.

The styling gained some mixed reactions, but the V8 power-unit was highly regarded by everyone who experienced it. Writing in the *Autocar* of September 1959, one driver was said to have remarked, *"Whether I were to choose the new Daimler or not, I should always be glad to have its engine in any car."*

The SP250 was displayed at the 1959 Earls Court Motor Show, and Hooper exhibited a close coupled saloon version of the sports car. Chassis no. 100571 was used for the prototype, which had a two door four seater body in metal. The standard features included wind-up side windows, hinged quarter lights, bucket front seats with hinged backs, a fully upholstered rear bench seat, and a leather padded instrument board. A heater/demister unit was also provided, as was a screen washer system and radio.

Unfortunately, the Hooper-bodied SP250 was not a success with the public, and this was probably the last Daimler ever built with a Hooper body, as the London works closed down later on in 1959. Perhaps Hoopers, more than any other coachbuilder, helped to contribute to the Daimler tradition, and it was indeed a sad day when the firm announced that they were going to bring production to an end.

When Hoopers finally closed the works gates for everything but servicing and repair work, their former Chief Designer, Osmond Rivers, spent several months at the Daimler factory in Coventry. Under his supervision, about five or six further bodies were built with a view to series production, although this time, glassfibre was used. A number of improvements to the suspension and steering were devised, and the interior layout was also redesigned slightly. After all this effort, it was decided in the end not to develop the model any further, and none of the cars are thought to have survived.

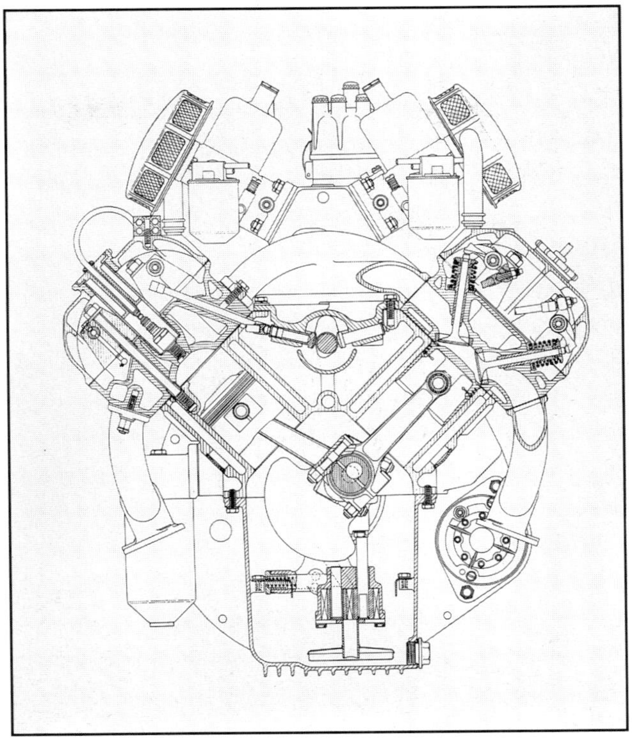

Above:
A cutaway drawing of the 2,548cc Daimler V8 engine.

The big V8s

Another new, but more successful model, was also shown at Earls Court. This vehicle, the DQ450 (or Majestic Major) had a chassis which was in fact a direct descendant of the pre-war New Fifteen. It had been this model to which Daimler had first fitted an independent front suspension, and although the New Fifteen was modified over the years that followed, the resemblance between the design and layout of the 1937 car and the Majestic Major were unmistakable. Similarly, the DB18s, Regencys and their descendants, all show a family likeness. Even the 9ft 6ins wheelbase of the Fifteen was retained, although the track gradually increased to accommodate the larger bodies of the later cars.

The Majestic Major was very similar to the 3.8litre Majestic, but there was one outstanding feature which set this model apart from its predecessor - an over-square 4.5litre V8 engine. Not surprisingly, the engine was built along much the same lines as its smaller 2.5litre brother, as Edward Turner was again the man responsible for this superb power unit. The engine had a bore and stroke of 95.25mm x 80mm, giving it a cubic capacity of 4,561cc. Cast iron was used for the block which had five main bearings, and once again, the single camshaft was positioned high up in the centre of the vee. The aluminium alloy heads featured hemispherical combustion chambers, and light alloys were used wherever possible. Two SU HD8 carburettors were used, and with a compression ratio of 8:1, the engine developed a healthy 220bhp at 5,500rpm. Several people have stated that the output was nearer 270bhp, but Daimler's dynamometer only registered up to 220bhp, so that was the figure quoted! In accordance with Daimler practice of earlier years, all the V8 engines were bench tested for a total of six hours prior to fitting.

External differences between the Majestic and Majestic Major were few, and concentrated on the frontal area of the car. The Major had a flying D fitted to the top of the radiator shell, and air intakes both sides of the radiator. These were circular apertures for the heater/demister unit and the chrome-plated grille covers featured a large vee, proclaiming the engine configuration. An extra 6 inches had been added to the cars length, most of which became boot space, and the rear bumpers were also extended. The Majestic Major was a very quick machine indeed, and for those who only managed to get a glimpse of the rear end of the car as it overtook them, the twin exhaust pipes should have offered conclusive proof as to which model it was.

Inside, the Major and the strict Majestic were very much the same. The larger model gained a tachometer and, in a bid to prevent glare and reflections at night, all dial surrounds were finished in matt black instead of chromium plate. When series production of

Below: *The SP250 prototype testing in Wales. The driver is John Box, who is currently building his own vehicle called the Tempest.*

Left:
An SP250 prototype. Thankfully the lines were toned down before the car went into production.

Right:
The SP250, or the Dart. Although the styling was not to everyones taste, there was no-one who could deny that the car had some excellent sports car qualities. This picture was taken in September 1959.

Below: *The Daimler 2.5 litre engine as used in the SP250, and (right) V8 engines on test prior to fitting.*

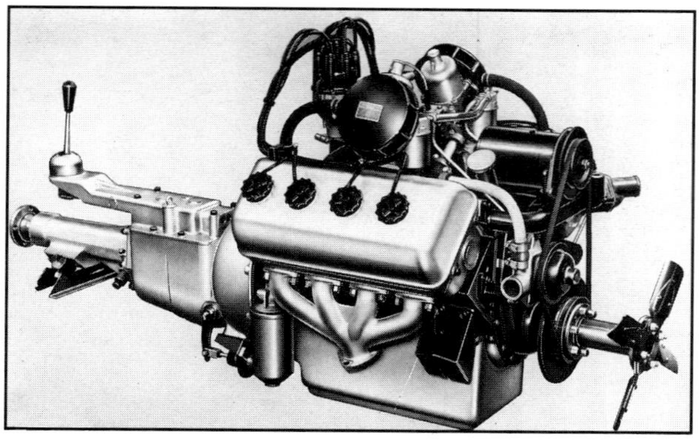

the car started at the end of the following year, servo-assisted disc brakes were fitted as standard, and a small warning light positioned in front of the driver had the dual purpose of telling him that either the brake fluid was getting low, or that the handbrake had been applied.

A prototype DR450 (the Majestic Major limousine, thought up by the Company's Director and General Manager, William Winters), was also on the Daimler stand. This had actually been built on a slightly extended 3.8-litre Majestic chassis by the Coventry-based Carbodies company and, it has to be said, was not particularly elegant. However, by the time the car was put into production about two years later, the styling had been significantly improved by Osmond Rivers, formerly the Chief Designer at Hoopers.

The Take Over

In 1960, still under Jack Sangster, the BSA Group managed to reap record profits of some £3.4 million, but at Daimler, the picture was not so rosy. The military contracts for the Ferret had dwindled, bus production was down to an average of only three units a week, and car sales were low due to a limited range.

In addition, Daimler had lost its Royal patronage - the end of yet another important era for the Company. A number of reasons have been put forward over the years for this fall from favour, but there was never any formal statement or any official explanation. It is probably that the Docker affair was partly responsible, but certainly not wholly. Perhaps the Daimler Company had rested on its laurels for too long, and the Royal household decided it was time for a change, taking advantage of the superior Rolls-Royce Phantom IV chassis, built solely for Royalty and Heads of State.

Anyway, at the beginning of 1960, Sangster was approached by Sir William Lyons of Jaguar, who was once again having problems with a lack of space. He enquired about the possibility of purchasing the Radford works, as the former Daimler Shadow Factory at Browns Lane was literally bursting at the seams, and unless he could find a suitable existing works nearby, he would have been forced by the Government into moving to a "development area", almost certainly miles away from his operating base.

The BSA Board considered Sir William's proposal and finally offered him the Daimler Motor Company as a whole. Sir William agreed and bought the Daimler name and business for the sum of £3.4 million. Of his negotiations with the BSA Chairman, he is later reputed

Below: *A Hooper bodied sports saloon based on the SP250 chassis.*

Above:
The Majestic Major limousine prototype. Note the standard Majestic front end. Air intakes would be fitted on either side of the radiator grille, with a stylized flying "D" mounted at the top of the fluted surround itself on all production Majors.

to have said, *"I do not recall a more amicable deal with anyone."* This feeling of geniality was undoubtedly helped when the two of them tossed a coin over some £10,000 worth of owed pensions, and Lyons won!

After the initial announcement made by Jaguar at the end of May 1960, that a deal was taking place and that they were soon to double in size, a second statement was released immediately after the formal proceedings were concluded, on 18th June 1960. The second statement ended: *"...... Jaguar Cars Ltd wish to deny unfounded rumours to the effect that sweeping changes, including even the extinction of the Daimler marque, are to be expected. Whilst one of the most obvious and immediate benefits accruing to Jaguar as a result of their purchase is the availability of much needed additional production facilities for Jaguar cars, the Company's long term view envisages not merely the retention of the Daimler marque, but the expansion of its markets at home and overseas."*

So Daimler was safe, but what was the deal with BSA. Jaguar's Board Minutes from 4th July 1960 give a clearer picture:

"1. The Chairman reported to the Board that pursuant to the discussions he had entered into with Members of the Board, he had signed the Agreement to purchase the Share Capital of The Daimler Co. Limited from The BSA Co. Limited, on the following terms;

a) Jaguar Cars Limited would purchase the 1,400,000 5% Preference Shares of The Daimler Company Limited for the sum of £1,225,000 (being 17/6d per Preference Share), and the 100,000 Ordinary Shares for £1,885,000, a total of £3,110,000 and,

b) Jaguar Cars Limited would also pay for the excess of current assets over current liabilities of The Daimler Company Limited, at a price to be jointly ascertained and certified by the Auditors of Jaguar Cars Limited and The BSA Co. Limited. It was resolved that the action of the Chairman in signing the said Agreement for the purchase be and is hereby approved.

2. It was reported the purchase was completed at 11.00am on Saturday the 18th June 1960, when the said sums of £1,225,000 and £1,885,000 making a total sum of £3,110,000 had been paid out of moneys standing to the credit of the Company's Accounts, but that the amount payable for the balance of current assets over liabilities had not yet been confirmed.

3. BSA Subsidiary Co's connected with The Daimler Company Limited.
It was reported that certain subsidiary companies of The BSA Co. Limited could be acquired, and IT WAS RESOLVED that the Chairman be authorised to conclude negotiations for the acquisition of the following:-

Daimler Corporation Incorporated
Transport Vehicles (Daimler) Limited
Lanchester Motor Co. Limited
Hooper & Co. (Coachbuilders) Limited
Barker & Co. (Coachbuilders) Limited

The total sum, as we know, came to £3.4 million, but for that, Jaguar had not only got a fully equipped factory with highly trained staff, but also a diversified business that had the potential to thrive under a great leader like Sir William.

BSA looked at the sale in a different way. In their 1960 Annual Report, it was stated by their Chairman:

Right:
The Daimler works in the Radford area of Coventry. Note the large area at the rear of the factory - this is the sports field where the old Radford Aerodrome was situated.

"I am sure Stockholders would wish me to comment on the circumstances leading up to the sale of the Daimler Motor Co. Ltd to Jaguar Cars Ltd.

"During the post-war years, the Daimler Company has on balance incurred a substantial loss, and I believe it is true to say that it could not have survived but for the financial support of the BSA Group. More recently, however, due to the technical progress made under the able direction of Mr Edward Turner, the position improved and I was able to report that moderate profits were earned in 1958 and 1959.

"When at the beginning of the year the new designs we had developed went into production on a comparatively small scale, the commercial outlook was not unpromising. Our manufacturing equipment, however, was not geared to quantity production and the additional plant and tooling necessary to produce in economic quantities would have required the investment of several million pounds. Recognising this position, your Directors were concerned as to whether acceptance of the risks involved in an investment of this nature and magnitude, despite the merits of our new designs, would be justified in the face of the vast expansion plans of the Motor Car Industry.

"Another factor adversely affecting future prospects, which would have had to be taken into consideration in arriving at a policy decision, was that the manufacture of military vehicles, which had occupied a significant part of our manufacturing capacity for many years, would continue only for a further limited period.

"These were the problems we faced and I am sure Stockholders will give unqualified approval to the decision taken by the Board when the opportunity arose to sell the Daimler Co. Ltd to Jaguar Cars Ltd."

As a matter of interest, BSA still had an eye on the car market, and brought out a three-wheeler microcar called the "Ladybird". Sketched by Edward Turner, it used the Triumph Tigress engine, and would have sold for £283 in the 1962 season. However, only two prototypes were ever built (at Carbodies) and the project was cancelled, this type of vehicle having fallen from fashion.

Chapter Eleven

THE HISTORY OF A LEGEND

For the purposes of this book, it is necessary only to briefly outline the history of Jaguar Cars Limited. The story starts with William Lyons who was born in Blackpool in September 1901. He left school at the age of 17 and was almost apprenticed at the Vickers shipyards, but went on to join the Crossley Motors outfit instead after his father managed to get him a job there as a trainee.

The young Lyons soon made his way back home, however, and helped in his father's piano restoration business. In his spare time, he kept up his interest in motorcycles and with three of his friends, reformed the Blackpool, Flyde and District Motor Club. It was through this Club that he met Jack Mallalieu, the manager of the local Sunbeam motorcycle franchise, and when Mallalieu asked Lyons to join him, there was little hesitation.

Facing page: *Sir William Lyons, founder of Jaguar Cars and preserver of the Daimler marque.*

Above: *One of the Swallow sidecars from the 1920s.*

In June 1921, William Warmsley, the son of a Cheshire coal merchant, moved into a house just a few hundred yards away from the Lyons' home. In his back garden, Warmsley began building sidecars for his friends, and eventually started building his Swallow combinations based on ex-WD Triumphs at a rate of about one a week. William Lyons approached Warmsley to buy a sidecar for himself, and suggested they should go into business together, producing sidecars on a much larger scale. After many late night discussions, it was eventually agreed that they should give it a try. Lyons left his old job, and the new partners eagerly set about their task.

When Lyons reached the age of 21, the Swallow Sidecar Company was officially registered. In a short space of time, extra premises were needed, but in the meantime, Swallows were doing well in competition, and several motorcycle manufacturers began to exhibit the sidecars on their stands at the major shows.

Warmsley's father, Thomas, purchased a large building in Cocker Street, on the northern side of the town and then leased it to what was about to be known as the Swallow Sidecar and Coach Building Company. The move meant that instead of having the business spread over three sites, everything could take place under one roof. A paint and trim-shop was set up in the new Cocker Street building, and a skilled workforce was obtained through placing advertisements in Blackpool and the Midlands.

Cyril Holland was one of the people recruited from the Midlands. Scribbling on a piece of paper one day, he found himself drawing a small two seater car. Holland was asked to do the drawing properly, and to make it suit the dimensions of an Austin Seven chassis - the first Austin Swallow was ready in the Spring. It was not that long before production was up to 14 cars a week, with the sidecars making a useful sideline. When Henlys of London put in an order for 500 cars, the decision to expand still further could be deferred no longer. During the autumn of 1928, the Cocker Street works were cleared out, and a new home for the company was set up on the Whitmore Park Estate in Coventry during November 1928.

Above: *The supremely elegant SS1 saloon, powered by a Standard engine.*

Below: *Completed Austin Swallows awaiting delivery from the new works in Holbrooks, Coventry.*

Above: *This beautiful 1938 SS100 sports car (chassis no. 39001) was destined for Prince Michael of Rumania. This was the first of the 3 ½ litre SS100s, of which only about 120 were built.*

Below: *The XK engine with its designers. From left to right; Harry Mundy, Walter Hassan, William Heynes and Claude Baily.*

Above: *The Jaguar XK120 fixed head coupé. The XK range of cars ran from 1948 to 1961, and were some of the most glamorous post-war machines ever built.*

Swallow went on to build on several different chassis; Austin, Morris, Fiat, Alvis, Wolseley and Swift among them, but probably the most important connection in those early days was the one made with the Standard Motor Company.

Standard running gear was used for the first SS cars - the SS1 and SS2, both introduced towards the end of 1931. Sales went well, and in a little over a year several overseas agents were appointed, many of them arranged by the Standard Company. SS Cars Ltd, which had been registered as a Swallow subsidiary late in 1933, was made into a public company in January 1935, with Lyons in sole command after the resignation of Warmsley. An engineering department was quickly set up and the Swallow Coachbuilding Co. (1935) Ltd, was formed to deal with the sidecar part of the business. From now on, SS would be officially recognised as the dominant marque of the company. After some careful thought about the name of his product, Lyons decided to adopt that of "Jaguar", and the new SS Jaguar range was launched in September 1935.

When the Second World War came, Coventry was hit particularly hard. The factory did as much as they could, manufacturing parts for aircraft, as well as carrying out service and repair work. During the war, the last links with Standard were severed, when SS Cars Ltd bought all the machine tools for the production of the Standard 2.7litre and 3.5litre engines. By the end of the war, the new name Jaguar Cars Limited had been adopted, and for William Lyons, a dream was realised.

F.R.W. 'Lofty' England joined the Company in the latter part of 1946. He had known Walter Hassan through his earlier days at Brooklands, and had served his apprenticeship at the London depot of the Daimler Company. He had also co-driven the Windovers-bodied Daimler Double-Six which took second place in the 1932 RAC Rally. Like Walter Hassan, who had joined the Company in 1938, he was to become an extremely important figure in the Jaguar story.

Fire destroyed part of the Jaguar factory in 1947, but the Mark V saloon appeared nevertheless,

Above: *One of the immensely successful Jaguar C types, on this occasion being driven by Duncan Hamilton during the 1953 Goodwood Nine Hour race.*

Below: *A 3.8 litre Mark IX Jaguar, built between 1959 and 1961. Around 10,000 were produced during this time.*

Above: *The Jaguar Mk.II compact saloon. This model was put into production in October 1959 and continued until it was replaced by a slightly modified version late in 1967.*

and more importantly, so did the twin-cam XK engine which would power Jaguar cars for many years to come. The public first saw the engine in four cylinder guise, fitted to the Gardner Special which took records on a Belgian motorway in 1948. At the Earls Court Show in 1948, the beautiful XK120 made its debut, sporting the now familiar six cylinder twin overhead camshaft XK engine.

Although not originally designed for racing, the XK unit was almost certainly flexible enough to be adopted for that purpose. Ron Sutton took a white XK120 to over 130mph at Jabbeke in Belgium, and not long afterwards, Jaguar's illustrious racing era began.

After the launch of the Mk.VII, Jaguar once again needed to expand. Lyons failed to get permission to build between the existing factory and Beake Avenue, so another major move was deemed necessary. Negotiations began to lease the No.2 Shadow Factory owned by Daimler at Browns Lane, Allesley. By the end of 1951, most of the equipment at Swallow Road had been successfully transferred to the new site - a site which Jaguar still occupies to this day.

The 1950s were dominated by the motorsport exploits of the Company. They took no less than five victories at Le Mans with the C and D type Jaguars, and in the 1956 New Years Honour's List, William Lyons was made a Knight Bachelor. During the following year, the decision was made to suspend the racing programme due to the newly introduced regulations, and to concentrate on the saloons instead.

Late in 1959, the Mk.II was launched. This was an improved version of the retrospectively named "Mk.I", which had been introduced about four years earlier. The first of the real Jaguar-Daimlers was based on this car.

Jaguar purchased the Daimler business from BSA in June 1960. The Daimler factory at Radford was part of this deal, and this allowed Jaguar production facilities to expand substantially. The plan was to move all car production to Browns Lane, while retaining the machining, engine-building, commercial and military vehicle operations at the Radford plant.

CHAPTER TWELVE

THE TRADITION CONTINUES

As they had promised, Jaguar carefully assessed all Daimler projects which were then in train, regardless of whether they were actually in production or at the prototype stage. The smooth transformation from Daimler to Jaguar was doubtless helped by some of the key staff staying in place, such as Cyril Simpson (who saw the V8s into production before retiring), and Dr Tait.

Dr Noel Tait is another of those unsung heroes who did so much behind the scenes work for the Company. Born in 1904, he studied Engineering Science at Liverpool University, and joined Daimler in 1926. Within three years he had established his own experimental department, becoming Assistant Chief Engineer to Simpson just before the Second World War. After the 1960 take-over, he was appointed the Group Engineering Consultant for the entire Jaguar combine.

Facing page: *A Daimler Majestic Major limousine - DR450/1 - at Holyrood House, Edinburgh.*

1861
SF 1

Above:
The 1960 Earls Court Motor Show. On the Daimler stand, a two tone Majestic, an SP250 and a Majestic Major saloon nearest the camera. The Major went into full scale production shortly after the Show.

Right:
One of the earlier Jaguar built SP250s.

Dr Tait, described as a fantastic engineer by all those who knew him, died in 1993, and Cyril Simpson died in December 1978. Simpson's obituary stated, *"He was a great engineer whose work was much respected"*, and indeed, the list of projects that he looked after for Daimler stand as a testament to his skill.

Edward Turner was another man who stayed behind, although mainly on a consultancy basis. However, Turner's strong personality clashed with the temperament of Sir William Lyons, and their relationship was very stormy. Retiring in 1964, he died at the age of 72, but will always be remembered throughout the automotive and motorcycle industry as an engineering genius and an astute businessman.

The Majestic had already received several minor modifications to its front suspension, mainly concentrated around the car's shock-absorbers. Daimler decided that these changes warranted a new chassis designation, so type number DF318 was given to right-hand drive cars, and DF319 was the number applied to left-hand drive vehicles. The DF318/9 also featured a higher rear axle ratio (earlier cars had been fitted with a 3.92:1 axle, but on the newer models this was changed to 3.77:1), and the tyre size was increased slightly. In addition, there were also minor alterations to the rear brakes and a revised instrument layout. With these modifications, the Majestic was probably at the height of its development, so Jaguar left well alone. They did, however, bring in the option of powered-steering costing around £100.

Production of the SP250 sports car continued as before, although most of the Jaguar engineers were unimpressed by it. Work began straight away on trying to improve the car, and it was decided that above all it needed a more rigid chassis and bodyshell. The B Specification SP250 was announced in April 1961, and boasted a new three piece channel section steel hoop around the bulkhead, reinforced metal B posts and bolt on longitudinal chassis outriggers. These modifications were not all implemented at once, however, and cars can be found with only certain changes to the chassis. Front and rear bumpers were also introduced as standard on the B Spec models, and a number of minor improvements were made in the cockpit such as redesigned seats and the added bonus of an adjustable steering column. A reserve petrol facility was now fitted, as was a windscreen washing system and exhaust tail-pipe finishers.

The Major

The 4.5litre Majestic Major was put into series production during November 1960. The Major was introduced at £3,083 including purchase tax, and powered-steering was offered as an option at £66; four years later, this became a standard fitment. Very little had changed on this model since it had first been shown to the public at Earls Court in the year

Above: *A Jaguar built Daimler Majestic. These later cars, from about chassis no. 99300 (rhd) were given a new designation - DF318/9.*

Below: *Both of the 4 ½ litre Daimlers were much the same when viewed from the front, though the rear end was slightly different as can be seen from the photograph.*

Above: *Production of the 4 ½ itre Majestic Major commenced in November 1960, only months after the Jaguar takeover. Note the distinguishing features of the larger V8 engined model such as the flying D on the radiator, and the air intakes on either side of the grille surround.*

Below: *The rear compartment of the DR450 limousine.*

Above: *A shot of the interior of the Daimler SP250 sports car. It is interesting to note how diverse the marques products were at this time, especially when comparing the SP interior to that of the limousine.*

before. The limousine version, the DR450, was still a long way from being ready, and didn't appear until September 1961.

The production DR450 was basically an extended Majestic Major, and its styling was a vast improvement on the prototype. The limousine had an extra 2 feet of structure built between the front and rear compartments of the car, which were divided by an electrically or manually operated interior partition. The interior was much the same as the DQ450 in the front of the car, but the rear compartment could be finished to the customer's choice, either in leather, West of England cloth or Bedford cord. Two fold-away occasional seats were provided, so that five passengers could be carried in the rear compartment, with room for another three in the front.

At this time, there were only three British limousines readily available - the Austin Princess by Vanden Plas, the Rolls-Royce Phantom V, and the DR450. As regards performance, the Daimler was in a class of its own, and it was reasonably priced at around £3,500 including tax. By special arrangement, customers were also given the option of buying the DR450 chassis only (for building hearses) although at £1,899 few were ever sold in this form.

"Powerful and lively", the Major, according to the *Autocar*, was *"in a category of its own"*, having *"much to offer that is not found elsewhere."* On average about five Majestic Major saloons were produced each week for several years, and four to five DR450 limousines. Unfortunately, the original 3.8litre six cylinder Majestic saloon was to be a short lived model. It was phased out towards the end of 1962, when a totally new model took its place on the market.

Commercial Vehicles

The Fighting Vehicle department was still building the Ferret (introduced in 1951) but the most important Daimler commercial vehicle introduced during this period was the rear-engined Fleetline double-decker, which also appeared in 1951. A few single-deckers were supplied, but the model was always intended to be a 78 seater, with Gardner diesel power and an electro-pneumatic Diamatic transmission. Diamatic transmission was first marketed in 1957.

Below: *A Daimler Fleetline bus at the 1963 Zurich Trade Fair - it had made its debut in September 1960.*

Above:
By adding a V8 engine, and a few mainly cosmetic modifications, the Mk.II Jaguar was transformed into a new Daimler model.

A prototype single-decker (the SRO) was constructed during 1962, and a prototype Daimler truck was also built. Unfortunately, nothing came of either project, but considering Jaguar had bought the Wolverhampton firm of Guy Motors the year before, there would have been little point in producing another range of trucks to compete with one another.

The next new model was the Roadliner - a 50 seater with a 9.6litre Cummins V6 diesel engine capable of nearly 80mph. No Daimler commercial engines were built after 1962, although the traditional half-cab buses continued to be available up until 1968. The later models were fitted with Cummins or Gardner units, sometimes linked to the optional Allison automatic transmission.

The first Jaguar-Daimler

Introduced at the 1962 Earls Court Motor Show, the "2½ litre V8" was the first Daimler saloon to be directly influenced by the Jaguar take-over. The vehicle had the body of a Mk.II Jaguar, but with the famous Daimler fluted grille and boot plinth. This first piece of beautifully executed badge engineering was not fully Jaguar, however; as the small Daimler V8 engine was retained. After realising the potential of Turner's V8 engine, Jaguar had tested a standard SP250 unit in the factory Mk.I hack. The results were impressive enough to convince the engineers to build a production model, and eventually the V8 was shoehorned under the bonnet of the Mk.II.

Below: *The engine bay of the 2 ½ litre saloon. Although it looks a little cramped, day to day maintenance was in fact quite simple.*

Wheel rim embellishers were fitted on the 2½ litre V8 as standard for the first time ever on a compact model, and the hub caps carried the famous D emblem in their centres. To the rear, only the fluted boot plinth and discreet badges on the boot and bumper showed that the car was a Daimler model.

Above:
The Daimler 2 ½ litre launch convention, held at the Jaguar works. It's official launch was scheduled for the 1962 Earls Court Motor Show.

Right:
Another debutant at the 1962 Motor Show was this elegant 2 ½ litre Daimler V8 powered Ogle SX250. Unfortunatley, the car was never put into production as a Daimler, but it did form the basis of the Reliant Scimitar.

The 2½ litre V8's engine had a bore of 76.2mm and stroke of 69.85mm, giving a cubic capacity of 2,548cc. The valves were operated by pushrods via a single central chain-driven camshaft. Fuel was managed by twin SU HD6 carburettors fitted with a manual choke, and an SU electric fuel pump kept the petrol moving towards them.

Several minor modifications were needed; a new sump was manufactured to clear the Mk.II's crossmember, the water pump was repositioned, and a viscous fan was adopted which gradually slipped as engine speed rose, thus keeping noise down to a minimum. The fixing bolts used to hold down the cylinder heads were redesigned to enable removal while the engine was still in place, and the exhaust manifolds were also redesigned, along with new down pipes to allow the necessary clearance in the tight engine bay. The twin exhausts differed from the Jaguar model by emerging at the two rear corners after running the length of the car.

The 140bhp engine was coupled to a Borg Warner Type 35 automatic gearbox, though the unit used in the Daimler lacked the intermediate gear hold which its Jaguar brother enjoyed. Initially, a hypoid bevel rear axle with a ratio of 4.55:1 was used, but this was later changed as it was considered to be too low. Dunlop disc brakes were employed all round, and gained acclaim from everyone who used them.

The independent front suspension consisted of coil springs and wishbone, with enclosed telescopic hydraulic shock absorbers and an anti-roll bar. To the rear, cantilever leaf springs were used, mounted in Metalastik blocks. Lateral location was by Panhard rod, fore and aft by radius arms. As the V8 engine was far lighter than any of the equivalent Jaguar power-units, the front spring ratings were adjusted accordingly, as were the shock absorbers. Steering was via a Burman re-circulating ball type unit, and the power-assisted steering option (again a Burman unit), incorporated an oil cooler.

The all steel monocoque body weighed in at 3,046lbs when dry (ie. complete with tools and jack, but without any oil, water or petrol), with the distribution 57% to the front, 43% to the rear.

The 2½ litre V8 was introduced to the public in October 1962, and was offered as a luxury alternative to the sportier Jaguar Mk.II range. The name of Daimler had always been associated with this corner of the market, so this sales strategy made a lot of sense. Although the 2.5litre engine could be no match (at least in performance terms) for the high-powered 3.4litre and 3.8litre XK units, the V8 was a lot smoother. At its first Motor Show, the car was awarded with the Gold Medal for coachwork - a fine start to the model's history. The award-winning V8 was accompanied Earls Court on Stand 98 by a Majestic Major saloon, a DR450 limousine, and an SP250 in B Specification trim.

The SX250

There was another Daimler based car at Earls Court in 1962 which attracted a lot of attention; the SX250, exhibited by David Ogle Ltd, and built on the Daimler SP250 sports car chassis. The Ogle SX250 was a very pretty car, styled in the mould of a sleek Grand Touring coupé. The body was manufactured in glassfibre, and was slightly larger than the SP. This particular show car was finished in twelve coats of sable brown cellulose and the interior was luxurious, with reclining seats, a radio and heater. Chrome wire wheels were included in the estimated price of around £2,500, and it looked as if David Ogle was on to a winner. However, only two cars were ever built and the project was abandoned, although the styling later formed the basis for the first version of the Reliant Scimitar.

2½ litre V8 News

The initial price of the 2½ litre V8 was quoted as being £1,298 list price (or £1,785 15s 3d including purchase tax), although within a month of the launch over £200 was taken off because of a

Below: *An interior view of the new 2½ litre saloon.*

Above: *The Radford works pictured during the period of change for the Daimler Company.*

reduction in purchase tax.

On introduction, the V8 model was more expensive than the 3.8 Mk.II by around £60. The range of options included seat belts, wire wheels, reclining front seats, powered-steering and a radio. A total of 12 colour options were available for the V8, and these (and the leather interior options) remained the same until mid-1965. Below is a listing of the schemes, coachwork options first, followed by the leather interior option:

Cream	*Red, light blue, dark blue.*
Mist grey	*Red, light blue, dark blue, grey.*
Sherwood green	*Suede green.*
Cotswold blue	*Light blue, dark blue, grey.*
Black	*Red, grey.*
Opalescent silver blue	*Dark blue, grey.*
Opalescent dark blue	*Light blue, grey, red.*
Opalescent silver grey	*Red, light blue, dark blue, grey.*
Opalescent gunmetal grey	*Red, light blue, dark blue, grey, beige.*
Opalescent dark green	*Suede green, beige.*
Opalescent maroon	*Maroon, beige.*
Golden sand	*Red, light tan.*

The first delivery figures for the V8 were released on the last day of July. Within nine months of its introduction, the 2½ litre V8 had made 821 sales on the home market, and 271 abroad.

SP250 News

The 1963 Motor Show was held at Earls Court from 16th to 26th October. Daimler occupied Stand 134, and showed the same range of vehicles as they had in the previous year, the only difference was that the SP250 was a C Specification model rather than a B. The only details which set the C Specification car apart from its earlier counterparts, was the addition of a heater/de-mister unit as standard, a cigarette lighter and a socket for a trickle charger. The SP250 won its Class in the coachwork competition for the second year in a row.

Peter Westbury won the 1963 RAC Hill-Climbing Championship in his Felday-Daimler. At the end of the 1962 season, he had finished seventh using a Daimler V8-powered Cooper, but during the autumn and winter which followed, Westbury set about building another machine capable of even better results. An SP250 2.5litre engine formed the basis of the new car, to which he added a simple 2inch SU carburettor and a Roots supercharger. This was linked to a Lotus five-speed gearbox, and the body looked similar to a Formula 2 car of the time. The Felday-Daimler was extremely successful, setting a new course record at the Craiganlet hill climb in Ireland, before finally going on to take the Championship at Dyrham Park in Gloucestershire.

More V8 Changes

Because of its introduction date, the V8 saloon naturally inherited many of the improvements already made to the Mk.II Jaguar. Among the first of the notable changes to the Daimler was the changing of the rear axle ratio from 4.55:1 to 4.27:1. This took place during January 1964, and made the car more economical at the expense of acceleration. However, with 155lb.ft. of torque on tap, the V8 saloon was still no slouch when compared with the competition.

During April 1964 (from chassis 1A 3235 onwards), the transmission was changed from the Borg Warner Type 35 unit to a D1/D2 type, which would perfectly suit the V8's characteristics. For those not familiar with D1/D2 transmission - when D1 is selected, all three ratios are engaged automatically; in D2, the lower gear is rendered inoperative. The dual-drive gearbox underwent a few slight modifications through its career, but stayed basically the same.

The end of the SP

At the Earls Court Motor Show of 1964, the Daimler stand exhibited only three models: the 2½ litre V8, the Majestic Major saloon, and the eight seater limousine based on the same vehicle - the Daimler SP250 sports car had been phased out during September.

The SP250 had never really been loved by the staff at Jaguar, but they did devote a lot of time and money in developing the car into something quite competitive. There were several experimental SPs built, including one fitted with rack and pinion steering taken from a TR4. This particular car also had a boxed chassis and a much improved axle location, but even though these

Below: *The Police found the qualities of the SP250 ideal for pursuit and motorway work.*

modifications made the car handle much better, they were never put into production. This experimental car was later bought and raced by Trevor Crisp, now Jaguar's Chief Engineer. Another SP250 was built with a Jaguar E type front suspension and coil spring independent rear suspension - again nothing came of the project. The same fate befell a marine version of the V8 engine, although a few were sold to the public.

By far the most interesting of the Jaguar experiments, was the Mk.II SP250 (or SP252). At least two of these cars were built - one fitted with rack and pinion steering - and they featured completely restyled bodywork. Sir William Lyons had a hand in the car's new lines, but on seeing the finished article, he was apparently unimpressed. The project died there and then, the cars were stripped down, and eventually one SP252 was sold. This vehicle was found again quite recently, and has since been restored.

The main obstacle in the development of the SP252 was that time spent on such a project meant less time was allocated to Jaguar's proven winners. In addition to this, all the necessary revisions to the Daimler would have pushed the SP's price into the E-type league, and to have an in house competitor was totally out of the question. The Daimler sports car just had to go.

If the small Tornado company had not failed, the SP250 might have lived on. Tornado had built a one-off Talisman on a shortened and strengthened SP250 chassis during 1963-64, and had approached Jaguar with their idea for taking the US market by storm. The Jaguar management had shown an interest in the venture, but unfortunately, the small Hertfordshire based firm ceased all car production after a six year run shortly afterwards.

Lotus Cars, still under Colin Chapman at the time, were also interested in the V8 engine for their Elan Plus 2. With a possible Jaguar-Lotus merger on the cards, this would have made a lot of sense, but in the end talks broke down with the Coventry firm, and began with Ford. Several engine options were discussed for the project, but eventually the now familiar twin-cam Ford-based unit was adopted.

All in all, it has to be said that the SP250 was a commercial failure. The gamble with the American market had not paid off, and only 1,201 SP250s were exported as lhd. Only 1,453 rhd cars were sold between 1959 and 1964, and when added together, this was well below the initial forecast of 7,500 sales in the first three years. Competition from cheaper cars with a proven sporting pedigree did not help, but the lack of sales Stateside may be put down to its image. Jack Wickes was once told whilst in the USA: *"You've designed an American-style car for the American market, but we want an English-style car!"* - apart from an under developed chassis on the early cars, I think this was really the crux of the SP250s sales problem.

The SP250 is now highly regarded as a

Below: *Peter Westbury - winner of the 1963 RAC Hillclimbing Championship with his Felday-Daimler. The 2 ½ litre V8 engine was ideally suited to competition wotk, thanks largely to its free revving nature. Westbury had used the Daimler power in a Cooper chassis for the 1962 season, and David Goode went on to use the same arrangement.*

Above:
Standard (well, fairly standard) SP250s were also campaigned on the tracks, and they are currently finding favour in historic races. An SP250 won the 1990 Monte Carlo Challenge.

collector's item, however, and has even prompted a Kent based firm to start reproducing the SP in limited numbers. Using an American Ford fuel injected engine (a shade ironic in view of 1989), the "SP290" should enter production by 1995.

Saloon Car News

The Majestic Major limousine gained its first Silver Medal for coachwork at the 1964 Show - further medals were awarded at the 1967 and 1968 events. By now, both of the 4½ litre cars were sold equipped with powered-steering as standard. In March 1965, the price of the 2½ litre V8 stood at £1,647 6s 3d, and remained that way for well over a year. Also the limited slip differential became listed as an option.

Autocar magazine tested the 2½ litre V8 again in May 1966. They noted that the small V8 was one of the quietest and smoothest engines they had encountered, and that it now had a well matched transmission. Top speed had increased slightly due to the higher ratios used in both the gearbox and rear axle, and although ultimate acceleration suffered slightly, fuel consumption was much improved.

Summing up, the *Autocar* stated;

"There is much in the finish, decor and equipment of a Jaguar Mk.II which must appeal to the discriminating tastes of the Daimler owner, and for those who have less need for a tiger under the bonnet, the substitution of the Daimler V8 offers extra refinement and better economy as compensations for the reduced performance."

A New Daimler

Towards the end of October 1966, the Motor Show was once again held at Earls Court. The 2½ litre V8 was there, along with the Majestic Major saloon and eight-seater limousine, but this year a new Jaguar-based Daimler made its debut - the Daimler Sovereign.

The Sovereign was the Daimler version of the Jaguar 420, which was a direct development of the S type. The S type had been introduced at Earls Court three years earlier as an interim model to fill the large gap between the Mk.II and Mk.X saloons, and had a modified Mk.II bodyshell, reworked to envelope Jaguar's independent rear suspension, first seen on the legendary E-type.

The design of the irs was based on that of

the double wishbone principle, though the top wishbone was actually a driveshaft, and the lower one evolved into a tubular forked arm which carried a large aluminium hub where the driveshaft was located. Trailing radius arms attached to the body floor and the lower arm kept the unit in check; like most of the components, these were rubber mounted. Two shock absorbers were employed on each side, each enclosed by coil springs. As four spring and damper assemblies were used, each could be made smaller, thus making the whole suspension a more compact unit.

Inboard disc brakes were used, these being attached to the differential, with the driveshafts running from them. Most of the unit was enclosed within a subframe, which acted as a mounting point for the differential, and again there was no metal to metal contact - the subframe being rubber mounted for noise insulation. This suspension would be a feature of all Jaguar and Daimler models for many years to come.

The overall styling of the vehicle gained some mixed reactions, but it was unmistakably a Jaguar design. The handling, however, because of the irs, was a marked improvement over the smaller Mk.II.

October 1966 saw the launch of the 420, an S type with a 4.2litre XK engine, and a restyled front end, very similar to that of the 420G (or Mk.X). The Daimler version was introduced at the same time and marketed as the Sovereign - badging and suitable fluting applied to the grille and boot lift identified the model. The 420 in Jaguar guise had a short life and was discontinued after only two years with the introduction of the XJ6; the Daimler Sovereign was allowed to run on for another year after that.

The irs was to become a trademark of future cars produced by the Company, but so too was the XK engine. The experimental work had begun on the XK engine during the latter part of the Second World War. William Heynes headed the team and, together with Claude Baily and Walter Hassan, developed this beautiful power-unit from the idea stage to production. Surely these three fine engineers could have never envisaged that their creation would stay rated as one of the world's greatest engines for so many years.

Above: *The small V8 engine was equally at home on the water, as this Daimler advert clearly proves.*

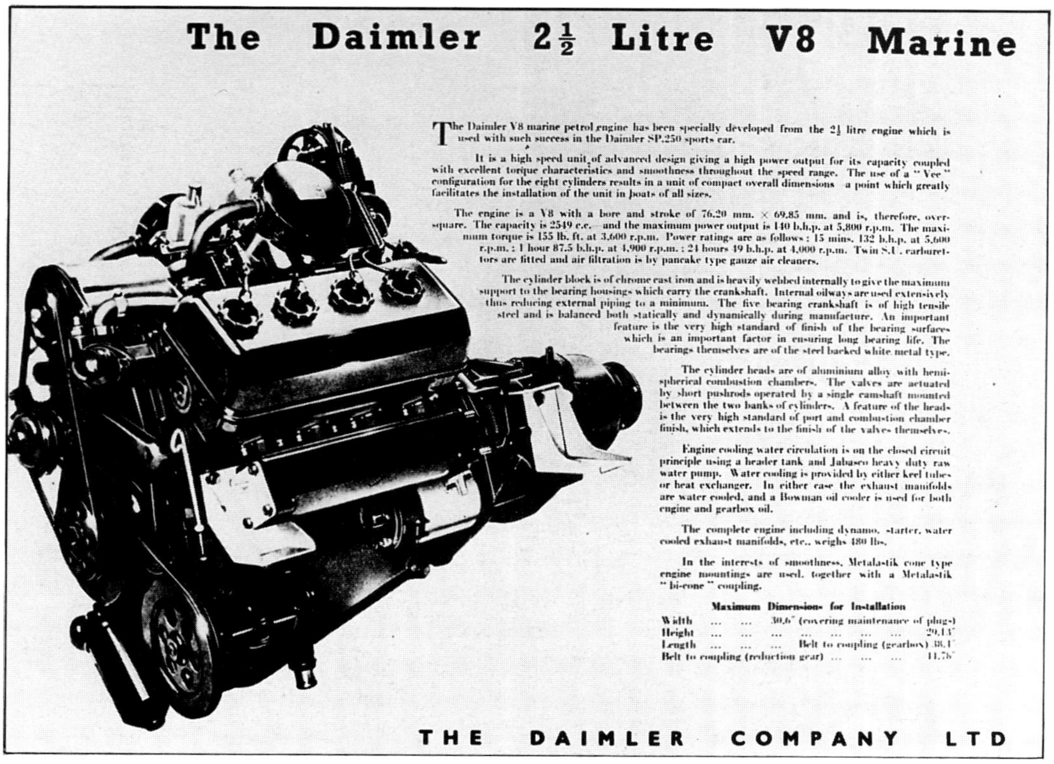

Left:
Craftsmen at Park Sheet Metal adding the extra length needed for the limousine model of the Majestic Major.

Below: *One highly successful Daimler powered boat was the Simmons Ski Boat, often seen on Lake Windermere in the hands of record breaker, Norman Buckley.*

Left:
The last SP250 ever built, photographed at Jaguar's Browns Lane works towards the end of 1964.

Right:
Sir William Lyons looking over the styling mock up of the proposed Daimler SP252.

A number of engine layouts were tried over the years and the true forerunner of the XK - Type XJ - was a 2 litre four cylinder unit. The twin overhead camshafts, hemispherical combustion chambers and cross-flow valve arrangement were all carried over. The XJ four cylinder was followed by an XJ six cylinder engine of 3.2litres, but this lacked low-speed torque. The stroke was increased, and the engine was termed the XK.

Twin overhead camshafts had always been associated with high performance machinery, but with it came expense - Duesenburgs and Bugattis have never been cheap, but Jaguar's extensive development and attention to detail kept production costs to a minimum, and value for money was a key selling point, as with all Jaguar products.

Many items were fitted as standard on the Sovereign, and included Varamatic powered-steering, heated rear window, alternator, reclining front seats, and a heater which allowed variable temperature control to the front and rear. If a manual gearchange was specified, overdrive came as standard, though close-ratio gears were available to special order. With the automatic gearbox, the type was changed from the old Model 8 to Borg Warner Type 35, because of the extra torque generated by the engine.

Performance was very impressive in its day; the 60mph from standstill yardstick shows it was capable of sub-ten second runs with either transmission (rather unusually, the model with the automatic gearbox was quicker). The manual Sovereign managed 0-100 mph in a shade over 27 seconds, while the automatic car achieved 0-100mph in 29.8 seconds - still pretty good for a car weighing approximately 3,700lbs. Manual Sovereigns could attain a top speed of 123mph (the automatic was limited to only 115mph), but both were nonetheless good enough to see off most of the foreign competition, let alone the British!

There is no trace of an optional equipment listing in either the first Sovereign price list or the first sales brochure; it can therefore only be assumed that the options were the same as those listed shortly afterwards in the next brochure, namely fog lamps and air conditioning.

Options listed common to both the V8 and the Sovereign saloons were the 980T radio (complete with a remote control aerial), front and rear seat belts, a laminated windscreen, Waso steering lock, and wire wheels - available either chrome-plated or silver painted. The optional extras for the small V8 only included power-assisted steering, a heated rear window and reclining front seats. The Sovereign was introduced at £2,120 16s 1d for the overdrive model, and £2,198 4s 10d for the version equipped with automatic transmission.

On 30th July 1967, the first delivery figures for the Sovereign were released; the overdrive model made 284 sales worldwide, 149 of these staying in the UK. The automatic version had 1,341 deliveries recorded against it on the home market, and a further 158 went abroad.

Further V8 Changes

On 3rd January 1967, a letter was sent out from the factory to all overseas distributors. The details of this letter were not disclosed until 9th March, however, when an announcement was made at the Geneva Show of the availability of a manual-overdrive four-speed syncromesh gearbox for the V8 saloon. The manual gearbox had been introduced for the V8 on the home market during February 1967 (on engine no. 7A 13231) once the engine mountings had been strengthened. The gearbox was available as either standard (this unit being the one used on the Jaguar 240) or with overdrive fitted. The overdrive worked only on the top gear, and gave a ratio of 0.78:1. On the few vehicles fitted with overdrive, the norm was to fit a rear axle with a

Right:
This special Majestic Major landaulette, built on the DR450 chassis, was produced for the King of Thailand.

Left:
Had it been less expensive to produce, this would have been the SP250 replacement - the SP252. Note the front indicators, straight off the Series I, E type Jaguar.

Right:
The prototype Daimler Sovereign. The official launch date was set for the Earls Court Motor Show.

4.55:1 ratio, which had been the original specification for the model when it was first introduced.

Unfortunately, the fitting of the Daimler 4.5litre V8 engine into the Mk.II body never got beyond the experimental stage. The big V8 had enough power to propel the largest of cars at a rapid rate, let alone a vehicle the size and weight of the Mk.II. Sadly, the expense of the necessary modifications put a sudden end to the project.

The Mk.II Jaguars were among the best-loved of all Sir William Lyons' creations - and still are to this day - but by now, the competition was beginning to catch up with the ageing design. What was needed was a vehicle to tide things over until the XJ was ready for its launch. It was decided to slightly modify the existing vehicle, and to keep the price down by the use of economy measures.

The resulting vehicles were designated the 240 (2.4litre XK engine) and the 340 (3.4litre XK engine). The prices for the cars ranged from just £1,364 to £1,422, but because of costs, the use of Ambla took the place of leather, and the option of the 3.8litre engine was no longer generally available. To coincide with the launch of the Jaguar 240/340 range in September 1967, a new Daimler model was introduced. This was a revised version of the 2½ litre V8, and was called the V8-250.

Distinguishing features included new slimmer bumpers and fog lights - the latter being a luxury lost by Jaguars due to the economy

Left:
The rear end styling of the 2 ½ litre V8 saloon.

Below: *An interesting picture from the press launch of the 420 range. From left to right: the Daimler Sovereign, the Jaguar 420G (basically the old Mk. X, which would go on to form the basis for a new Daimler limousine) and in the background, the Jaguar 420.*

Right:
The 1966 Sovereign. Notice how much heavier the car looks without the optional wire wheels. However, even with standard wheels, the car looks nimble enough on the move.

Left:
Introduced in August 1967, the V8-250 came about due to cost cutting measures deemed necessary across the whole Jaguar and Daimler range.

drive. Wheels and hub caps were, by now, standardised throughout the whole of the Jaguar and Daimler range, but rim embellishers were unique to the Daimler. Inside, reclining seats were now a standard fitment, and a padded roll finished in matt black was used on the dashboard, carried through to the door cappings as with the Sovereign. Leather continued to trim the seats, but the panels were now ventilated. An alternator and a heated rear backlight were standard fitments on this new small V8 saloon, and when the price list was released in October, prices ranged from £1,615 12s 4d for a manual version, to £1,697 19s 5d for an automatic.

A few months after V8-250 production started, separate air cleaners (one for each carburettor) were adopted, and there were some slight engine modifications. In March, there was a substantial price increase on the small V8, and opalescent paint colours were dropped from the range of standard schemes. At this time, the Majestic-Major saloon was priced at £2,958 including tax, and this figure was to stay the same until production of the model ceased a couple of months later.

Corporate Changes

In July 1966, Jaguar Cars Ltd had merged with the British Motor Corporation to form British Motor Holdings, with Sir William remaining in control of his company. In May 1968, however, after a merger between BMH and Leyland, Jaguar found themselves part of a new business known as the British Leyland Motor Company.

Due to this latest amalgamation, Daimler bus production was eventually transferred to the British Leyland Truck & Bus Division. In the meantime, development work continued, including the use of the 8.4litre Perkins V8 diesel as an optional power unit for the Roadliner, and the choice of either an AEC or Cummins engine in the CR36, which had a longitudinal rear engine location. After 1971, only the Fleetline remained in full production to keep the long tradition of Daimler commercials alive, and by the mid-1970s, it was being offered with the Leyland turbo-diesel engine. By this time, the vehicle was built at Leyland's Lancashire plant and, by 1979, it was actually being badged as the Leyland Fleetline.

So, the link between the Daimler Motor Company and commercial vehicles finally came to a rather abrupt end after over 80 years. By the mid-1960s, when the Fleetline had become firmly established, Daimler had become the world's largest producer of double-decker chassis, with something like 30 units being built each week - such figures serve as a sad reminder of a glorious past in this field.

Military vehicles, too, were gradually phased out - the Ferret finally being killed off in the early-1970s after its replacement was introduced in 1973. In the meantime though, the Fighting Vehicle section of the Company had been involved with some other armoured car projects, including the very machine that took the Ferret's place. Lofty England sets the scene:

"From the time we [Jaguar] *took over Daimler in 1960 and I became Assistant Managing Director, the Fighting Vehicle business came under my direct management. The fighting vehicle section of Daimler, ably led by Ted Wilday, was solely an engineering department with no administration or general management. So in effect I was both General Manager*

Below: *Bob Crouch (left) with the 1968 Roadliner coach which was sold to the Blue Bus Line. It was destroyed in a fire some eight years later.*

Above: *The Fleetline bus of 1968. It remained basically the same until production came to an end in 1973.*

Below: *The last Daimler bus chassis being delivered in 1973.*

and, since Bill Heynes did not want to be involved with AFV or bus, I was also Technical Director!

"In 1965, we were invited to a meeting at FVRDE Chobham to discuss the future FV requirements. The meeting was with MOD, Army top brass (the users) and companies making AFV's. Ted Wilday [Chief Engineer of the AFV division] and I attended.

"A presentation was made of the requirements and we were shown the ideas of FVRDE who had already made a running prototype of the tracked vehicle, and also a wooden mock-up of what they were suggesting for Fox/Vixen [the Ferret replacement], Fox to have a crew of two, and Vixen three. Their mock-up had three axles (six wheels) and they were also suggesting that both vehicles should use the 4.5 V8 Daimler engine, which was light in weight and had the right power output. We were asked to go away and then put up our proposals.

"So with the great experience of Ted Wilday, we got together and put up the two axle Fox/Vixen proposal and the use of the 4.2 Jaguar engine in both Scorpion and Fox, since in militarised form (which meant only a single downdraught carburettor and screened ignition), it was narrow enough to sit the driver beside the engine in the Scorpion. And, since that engine would continue in production for some years and parts were available from Jaguar distributors and dealers around the world, it was a better proposal than the Daimler V8 which had only been tooled up for an output of 40 per week and which we would not be continuing.

"Our suggestions for the Fox/Vixen were accepted, and we were given the contract to design, develop and

Above:
A V8-250 Daimler at the Royal Show, at Stoneleigh in Warwickshire.

Left:
Lord Brabazon of Tara with FLY 1, his Majestic Major saloon.

Above: *The Daimler Fox, developed by Jaguar in the late 1960s.*

Below: *The Fox was built to withstand water diches and a variant, the Vixen, was fully airportable. The Fighting Vehicles department at Daimler were still, for the time being at least, a very active unit.*

Right:
The latest prototype fighting vehicle in the Water Test tank to the rear of the Radford works.

produce 15 prototypes.

"This was done in conjunction with a project manager appointed by FVRDE and all carried out to a critical path plan set up by Ted Wilday, who was not only an excellent AFV engineer, but a good manager with the ability to get things done.

"The prototypes passed all their tests including tests carried out in the tough conditions of the Australian desert but, as is usual with MOD jobs, the production contract was put out to tender at a fixed price. I had heard before the Director of Contracts came to see me that we were not the lowest tenderer, so it was no surprise when he told me that, and I replied telling him who it was - namely ROF (Royal Ordnance Factory) Leeds, who wanted the work and would not be out of pocket if their tender lost money. He told me that Vixen would be the standard vehicle after the first 100 Fox vehicles had been built, but we could have the contract to build those. I thanked him but said I was not prepared to allocate factory space, install equipment and employ people who, after a few months, would no longer be required.

"However, as the Jaguar engine had proved satisfactory in both Fox and Scorpion, the Ministry could now negotiate a non-fixed price contract with me for the supply of those units, which was done. We installed special test cells and set up the production facilities which have since produced a large number of Ministry engines.

"Why the policy changed from Vixen to Fox since that time I have never heard. When this happened the Projects Manager of FVRDE wrote to me to thank us for the good job done by Ted Wilday and his team, and to say that in his opinion we had the best fighting vehicle engineering team in Britain."

After Ferret and the brief encounter with the Fox/Vixen project, the Company's involvement with fighting vehicles dropped off, although Jaguar continued to supply the military version of the XK engine to a number of manufacturers, including Alvis, the builders of the Scorpion tank mentioned by Mr England.

Mid-way through 1968, a new Daimler limousine was launched, marking the end of the DR450. By now, both the 4.5litre limousine and saloon had gained limited-slip differentials and alternators, but otherwise little had changed since they were first introduced. The Majestic Major saloon had already been dropped from the range, but the Majestic Major limousine had one final blaze of glory when it was awarded the Silver Medal for coachwork at Earls Court. This was to be the model's third and last such prize, as the DR450 was not exhibited again.

By the time the delivery figures were released at the end of September 1969, the last of the V8-250s had long since been driven off the production line. The small V8 had again sold well, as had the Sovereign, but both would be entered on the ledgers for the very last time. A new era of true Jaguar-Daimlers had begun.

CHAPTER THIRTEEN

THE DS420 LIMOUSINE

On 11th June 1968, the first model since the integration of Jaguar and BMC was officially announced. The DS420 would have coachwork by Vanden Plas (also part of British Leyland), and thus replace both the Daimler DR450 and VDP Princess limousine in a single stroke.

The DR450 had only been introduced in September 1961, but although the 4.5litre V8 engine was superb and - as one would expect - the level of luxury was very high, the car was let down by the archaic specification of its chassis. In addition, its looks, although full of character, would certainly never have stood the test of time in the same way as the DS420 has.

The Vanden Plas Princess on the other hand, had been put into production some ten years earlier, and only updated cosmetically since. By 1968, VDP were still trying to decide on a final shape for its latest styling revision, while still employing the old chassis. The old Austin 4-litre engine dated from the 1940s, and although the XK engine had been designed in the 1940s too, the Austin unit performed like a 1940s engine, developing a lazy 120bhp at very low revs.

Facing page: *The Daimler limousine, a 1990 model year DS420.*

G97 WHP

Above: *A prototype DS420 body - at this stage nothing more than a styling mock up. Note how much shorter it was than the production model, confirming reports that a close coupled saloon version was on the books.*

The DS420 limousine was aimed at the prestige sector of the market, where dignity of styling, individuality of appearance, and the highest standards of refinement and comfort were all essential characteristics. During the very early stages, the design team had considered building a close-coupled saloon version, probably along the lines of the very rare Jaguar Mk.X limousine, but unfortunately nothing came of the idea.

Built of steel, the body of the DS420 was based on unit-construction principles to provide the stiffness necessary for the full benefits of fully independent suspension to be obtained. On the Daimler, two massive fabricated box section side members were welded to the heavily corrugated floor to form the floor section. An extremely stiff scuttle assembly, also largely comprising of fabricated box-section members (to which the front end structure was welded), was welded to the front of the floor section.

The rear consisted of a fabricated assembly incorporating the rear transverse bulkhead, the luggage compartment floor and inner wing panels. Further box-section transverse members were added to provide additional stiffness at strategic points, and by these means, an immensely strong platform chassis was created to which the body structure and panels could be welded. Most of these sections were borrowed from the Jaguar 420G in order to keep the tooling costs to a minimum, the main difference being a 20 inch panel inserted in the floor to give the vehicle its extra length in the wheelbase.

The body was styled by Daimler to provide generous accommodation for seven or eight people, together with a very large luggage compartment. In fact, its nearest competitor in terms of luggage capacity was the Mercedes, which had only two thirds of the capacity of the Daimler. The regal styling incorporated a sculptured recess treatment of the waistline, which not only lightened the appearance of the body, but also provided a continuous harmonising feature between the front and rear end styling, all of which had more than a hint of Hooper inspiration in it.

The modern interpretation of the traditional Daimler grille provided a focal point for the front-end treatment, which incorporated the latest Lucas four-headlamp system and slimline bumpers - all very pleasing to the eye, and not dissimilar to the 420G. Styling at the rear resulted in a car of majestic proportions, allied to a grace and modernity which was already becoming increasingly rare in this type of vehicle.

All body panels were supplied by Motor Panels, with the Mk.X (420G) pressings coming from Pressed Steel Fisher. Contrary to popular belief, the vehicle was built at Park Sheet Metal (based at the northern end of Coventry) to the body-in-white stage. The cars were then transported to the Vanden Plas works four

Above:
The front compartment of the new DS420 limousine....

at a time, where they were treated and finished to a point ready for sale. Park Sheet eventually purchased the Mk.X jigs, which allowed the complete build operation to be carried out in-house with the aid of the larger jigs already there designed by Maurice Britain - one of the Company's future Directors.

The lavishly furnished interior reflected the same care and extensive use of the finest quality materials as Daimler and Vanden Plas models of the past. The rear compartment provided generous accommodation for six people; three on the rear seat, and three on the fold-away occasional seats. The rear seat in particular offered superlative standards of comfort, and was upholstered in either West of England cloth or high quality leathers selected by Connolly. The rear seat armrests, located at the sides, each housed an ashtray, cigar lighter and courtesy light switch - heating and ventilation controls were also incorporated

.....and (left) a standard rear compartment trimmed in West of England cloth.

Right:
The elegant lines of the new limousine had a distinct Hooper look, especially when viewed from the rear, despite no involvement by the London concern whatsoever

in one armrest. All cabinet work was executed in polished walnut, and the final touch of luxury was provided by the deep-pile fitted carpet which covered the entire floor area.

The rear compartment was isolated from the front by a fixed division with sliding glass, although an electric division could be specified at extra cost. The front compartment featured a bench seat for the driver and a passenger. A comprehensive set of instruments was fitted, similar to those of the Mk.II Jaguars.

However, the advanced technical specification was the outstanding feature of the new limousine, particularly in comparison with the two vehicles it replaced. The 4.2litre XK engine was used (as fitted to the Daimler Sovereign) which, with an 8:1 compression ratio and twin SU carburettors, developed 245bhp at 5,500rpm, and 283lb.ft. of torque at 3,750rpm. The V8 engines had been dropped under Leyland because, to quote Lord Stokes, *"they were hand-made, so too costly to produce."*

Borg Warner automatic transmission was fitted as standard. Fully-independent suspension and servo-assisted disc brakes were fitted on all four wheels, with independent hydraulic circuits to front and rear - features inherited from Jaguar. Power-assisted steering of the varying ratio type was fitted as standard equipment, giving a turning circle of approximately 46 feet.

Although the accent was placed firmly on appearance and luxury, the important safety factors

Above:
The rear compartment of the Congolese Ambassador's DS420, chassis no. 1M 1655. It had a roof mounted aerial for the television.

Right:
DS420 rolling chassis were sold to coachbuilders for conversion into hearses. This standard height example, dating from 1969 and built by Startins of Birmingham, is just one of many produce down the years.

Above:
A limousine from around 1973. Note the thin coachline along the front wing running backwards, which replaced the heavy chrome strip in January 1971. Note also the new rear quarter light arrangement, modified during the early part of 1972.

Above:
A low-liner version of the Startins hearse, circa 1974.

were not allowed to be overlooked. Independent hydraulic circuits to the front and rear brakes with tandem type servo-assistance, a universally jointed steering column, radial tyres, the immense strength of the monocoque bodyshell, and the highest standards of roadholding, steering and braking, all helped to make this a very safe car. Inside the vehicle, heavily padded windscreen rails and waist mouldings, and front and rear seat belt anchorage points (not obligatory at the time), all formed part of the standard safety features.

Three prototypes were built at Park Sheet Metal. The first two were trimmed and finished by Daimler themselves (chassis no. 1M 1001 and 1002) and the third went to Vanden Plas in London to be trimmed (chassis no. 1004), as did car number three, although this was the first drive-away chassis. Five of the six man team which built the prototypes still work at PSM, and the sixth now runs his own sports shop in nearby Bedworth.

This big 42cwt car, was certainly no slouch. The maximum speed was put at 110mph, and 0-50mph was achieved in 9.2 seconds, impressive enough by today's standards, let alone for a vehicle of such massive proportions over 25 years ago. Fuel consumption, not surprisingly, was heavy - between only 12 and 18mpg. After its launch at the Royal Lancaster Hotel in London (it was used as a practice run for the XJ6 launch), the limousine was introduced at £4,424 9s 5d for the standard car, and the drive away chassis was priced at £2,369.

There were only two standard coachwork colours, black and Carlton grey, with four optional colours of light grey, light green, dark green and maroon. The interior options were equally limited, standard trims were black leather in the front compartment matched with fawn West of England cloth to the rear, brown leather throughout, or blue/grey leather throughout. Alternative trims could be specified at extra cost, such as light brown, maroon or black leather throughout.

In August 1968, the first production commenced at Vanden Plas on car no. 1M 1005. The official list of options was included in the August price list. Many options were available and only the prices changed for many years to come. An extract from the August 1968 price list follows:

Radio to front compartment			*£54*	*6s*	*1d*
Radio to rear compartment			*£57*	*10s*	*0d*
Electric aerial (in lieu of manual)			*£31*	*18s*	*11d*
Electric front windows			*£63*	*5s*	*0d*
Electric rear windows/quarter-lights			*£125*	*17s*	*3d*
Electric glass division			*£121*	*7s*	*9d*
Chrome flag mast (to bonnet or wing)			*£8*	*18s*	*11d*
Electric clock to rear compartment			*£21*	*1s*	*8d*
Adjustable reading lamps (pair)			*£22*	*7s*	*3d*
Nylon rug to the rear			*£29*	*7s*	*9d*
Rubber mats to the front			*£5*	*15s*	*0d*
Wing mirrors			*£5*	*7s*	*4d*
Fog lamps (pair)			*£19*	*16s*	*1d*
Heated rear window			*£30*	*13s*	*4d*
Laminated windscreen			*£12*	*15s*	*6d*
Rear compartment footrests			*£19*	*3s*	*4d*
Air conditioning			*£319*	*8s*	*11d*
Seat belts -	*Front;*	*Static*	*£14*	*1s*	*1d*
		Inertia	*£20*	*8s*	*11d*
	Rear;	*Static*	*£15*	*6s*	*8d*
Non-standard exterior colour			*£57*	*10s*	*0d*

Right:
The Governor General of Jamaica took delivery of this superb landaulette, supposedly in preparation for the Royal visit of April 1975. Although similar vehicles were listed for quite some time as standard production models, officially only two DS420 landaulettes were ever built.

A duotone paint finish was also available, although this warranted a price on application status, as did the choice of fire extinguisher, due to there being so many different types. The prices shown are the total prices, including purchase tax.

In 1968, at the first Motor Show at Earls Court since the introduction of the DS420, there was a model on both the Daimler and Vanden Plas stands. The limousine continued to grace the Jaguar-Daimler stands into the early 1990s.

In February 1969, the prices were revised. The standard DS420 now retailed at £4,520 16s 5d, an increase of about £100. The price for the chassis only was retained. Two months later, a more substantial increase in price was incurred on the limousine - £4,874 12s 6d was the new asking price for a basic model, with the drive-away chassis priced at £2,419. A new option of a coat of arms plaque was listed at the same time. An illuminated plaque cost, £52 4s 6d, and the standard version was £39 3s 4d. Options such as Sundym glass were beginning to appear, and front seat belts would soon become a standard fitment.

The Daimler Hearse

In June 1969, a Daimler newsletter described the new DS420 hearse built by Thomas Startin Jnr of Birmingham. Built on the standard drive-away chassis (already fitted with the front wings, bonnet, bulkhead and front doors) was a frame of top quality hardwoods, mainly ash, the joints of which were then lead painted and strengthened with steel plates at stress points.

The large one-piece fixed side and rear windows were of plate glass, and the roof was mounted on bent ash bows - specially moulded rubber prevented water ingress. Body panelling, like the standard body parts, was in steel, but the back door was in 16 gauge aluminium to save weight. All panels were hand shaped, and painted twice on the inside with sound-deadening compound. The coachwork, designed by Ken Bilson, was built as low as was possible at the time.

The rails and window fillets were of french polished walnut, with the standard Daimler instrument panel and fascia being used. Two large and adjustable fold-forward front seats were fitted, and designed so that comfortable three-abreast seating was possible.

The first Startins hearse was sold to T.A. Ball of Birkenhead for the princely sum of £4,350. Sadly, Mr Ball himself was the first customer! A couple of years later, Startins built the low-line hearse in the same manner, and traditional hearses are still produced in much the same fashion to this day.

DS420 Sales

The first delivery figures for the DS420 (released in September 1969) were quite encouraging, though not up to the expected four to five units per week. Only 21 DS420s were sold abroad, and 182 at home. 26 drive-away chassis were also distributed, all within the UK. By the following September, when the next year-end reports were released, the DS420 had become established, with 374 cars sold on the home market and another 74 units going abroad. The demand for the drive-away chassis had doubled, with 57 vehicles being delivered.

Above:
An engineering masterpiece - the Jaguar XK engine. Only detail changes have affected the unit throughout its long life which began in 1948, continuing to be used to this day in certain military applications.

By December 1970, decimalization was widely accepted, but it could do nothing to hide the massive price increase of nearly 10% on the model range; £5,445.15 was the new asking price for the standard DS420, with the drive-away chassis now costing £2,702.00. Fortunately, after the massive increase of December, prices held fairly steady for several months.

The First Face-lift
The limousine had a minor face-lift in April 1972. The original two piece wind-up rear quarter-lights were replaced by single swivel-type windows, and the division's top frame was deleted, giving the rear occupants a feeling of more space and light. The chrome strip that ran along the waistline was replaced by a gold coloured coachline, and several trim modifications were introduced, all these changes taking place at chassis no. 1M 2180. By this time, the price of the limousine was £5,922.40.

In July 1973, the first of two landaulettes (chassis no. 1M 2488) was built - the second (chassis no. 1M 20099) was produced a year later. The first landaulette was built for the Governor General of Jamaica, and was used by HM the Queen and the Duke of Edinburgh on their tour of the country. Terry Phillips of Park Sheet Metal remembers the second car was finished in white, and was a left hand drive model, confirmed by the chassis number. Although listed as a standard model for some time, only two landaulettes are known to have been built. However, another DS has recently been converted into a landaulette by a DLOC enthusiast.

Vanden Plas
By 1974, Jaguar had taken over the responsibility of the Vanden Plas works. This famous coachbuilding company had originally started in Belgium as a family business, concerned with horse-drawn carriage components before eventually going on to build the complete article.

By the start of the twentieth century, the automobile was becoming more and more popular. Van den Plas was quick to realise the potential of this new form of transport, and thus offered a wide variety of body designs to suit various Continental chassis. It did not take long for the company to gain a favourable reputation, both for their flair in creating superb lines on paper, and for the quality of the finished vehicles.

The Van den Plas name came to England when Theo Masui, a designer of some repute, joined with Warwick Wright and a councillor of the Belgian Legation in London to form a company building bodies under licence from Belgium. This firm eventually became known as Vanden Plas (England) Ltd on the 6th May 1913.

During the First World War, Vanden Plas had to abandon its coachbuilding activities. The Company, as a subsidiary of the Aircraft Manufacturing Company (Airco), helped the war effort by making numerous aircraft components. Vanden Plas (England) 1917 Ltd emerged after the war from Airco, as did the de Havilland Aircraft Company.

Body building started again in March 1919, and the first Motor Show exhibits appeared in 1922. By the following year, Vanden Plas were in financial trouble and the Receiver was called in. Edwin Fox was the Manager of the Company at the time and he, along with his two brothers and a certain amount of outside help, made a successful bid to gain the Vanden Plas name. After renaming the firm Vanden Plas (England) 1923 Ltd, the Fox brothers went on to build up the now famous Kingsbury Works in London.

For the while the Company thrived under the control of Fox, building up an excellent relationship with the Bentley marque. It was not long, however, before they were in trouble again. Fortunately, the introduction of the Alvis Speed Twenty in 1932 saved them, as they built special bodywork to coincide with the models launch - many orders were received as a result of this.

With the advent of the Second World War, Vanden Plas once again served the country by carrying out aircraft-related work. Roland Fox, Edwin's son (b. 1911), joined the family business after gaining much of his training abroad. He also worked with various Midlands coachbuilders before he eventually settled down at Vanden Plas in 1940.

Just after the war, Austin approached VDP about building a quality saloon on their new six-cylinder chassis. Vanden Plas eventually became a subsidiary of Austin, the deal being concluded in June 1946, and six months later, the first product of this new partnership emerged. The Princess saloon was first shown at the Geneva Show, with the limousine version following in 1952 at Earls Court. The Princess name was later carried on several vehicles, including some of the badge-engineered Austins. The DS420 was the next major milestone in the firms history, and the Vanden Plas name was to be associated with Jaguar and Daimler for many years to come.

When the year's delivery figures were released in September 1975, not one export had been recorded against the DS420, and there was very little change on the home market. A total of 212 limousines had been sold, and 38 drive-away chassis. Six days after the figures were released, another large price increase took the price of the limousine over £109,000. Despite falling sales over the following years, limousine prices continued to go up in leaps and bounds. In June 1979, the £20,000 barrier was broken, and by the end of 1979, £23,546 was the asking price for the standard car. The drive-away chassis had also crept up to a hefty price, retailing at £14,064.

Another Face-lift

The limousine operation was moved entirely to Coventry during November 1979 following the closure of the Vanden Plas factory. The DS420 had by then undergone another major face-lift; neater square indicators and air intakes were used at the front, complementing the smaller overriders introduced a couple of years earlier. Sadly, the flying D on the radiator grille had disappeared, due to new safety regulations.

Prices continued to rise in June and November 1980, increasing the basic price of the standard limousine to £25,712.76. The familiar range of options was again offered, but the prices for these were also on the high side. Air-conditioning cost £2,425, and the Phillips

Right:
Smaller, neater over-riders were introduced for 1977, as shown on this vehicle supplied to HM the Queen Mother in the following year.

Above: *Limousine production as it used to be. The Vanden Plas works at Kingsbury produced over 2,500 Daimler limousines, but the factory closed during 1979. This is an early car, incedentally.*

AC880 radio/cassette was £868.46. The gearbox had been changed in July from the old Borg Warner units (Model 8 had first been used, with the Model 12 replacing it some years later) to the smoother GM400 automatic gearbox. The prices reflected in the delivery figures, which hit an all-time low; only 85 DS420s were sold in the UK, and 37 chassis were supplied. Exports were poor with cars going to Australia, Brunei, Nicaragua, Singapore and Hong Kong, but giving a year's total of only eight vehicles.

Another rise in February 1981 took the price of the basic limousine to £26,998. During November, however, a detailed appraisal of manufacturing costs and market research led to a revised pricing of the model. At this time, the standard Daimler was nearly £27,000. From 1977, the price of the DS420 had increased by 99%, and that of the chassis by 92%.

The DS420's main competitors at this time were Ford Granada-based vehicles by Coleman Milne, priced at £16,753 for the Minster, and £19,429 for the Grosvenor. To become more competitive, new vehicle prices and significant improvements in option costs were proposed by Keith Cambage, the Director of Limousine Operations. The new price for the limousine was to be £23,588.60, while the drive-away chassis was to be £14,110.50.

The basic specification of the limousine was to be changed to support the adoption of a more competitive pricing stance; front and rear compartment seats would be fully trimmed in vinyl - West of England cloth and leather becoming cost options. Front and rear carpets would be in Series 3 quality polyester pile in place of wool Wilton (although Wilton was used again later), and the boot would be trimmed with black painted and grained hardboard on the sides and back, with a PVC coated felt floor mat and spare wheel cover.

Standard body colours would be black, Carlton grey, dark green, dark blue, light grey and ivory, with an individual quotation supplied for any duo-tone, non-standard solid or metallic finish. Interior trim options included black, doeskin or blue/grey vinyl seat facings as standard, but also available (at an extra charge), was burnt umber and county tan, the last two being supplied with leather facings on the rear seats. In the extra cost range was black vinyl upholstery in the front with West of England cloth in the rear, Ratten carpet with headlining in fawn. Any non-listed trims, would again warrant an individual quotation.

It was hoped these new competitive option prices would give customers the expected standards of luxury at a more reasonable price. Optional extras were to include:

LW/MW radio to the front (push button, manual aerial)	*£149.50*
AM/FM radio/cassette to rear (electric aerial)	*£647.83*
Air-conditioning to the rear	*£1557.30*
Central door locking	*£274.08*
Electric clock to the rear	*£49.83*
Electric division	*£591.77*
Electric windows all-round	*£753.73*
Rear footrests in leather or cloth	*£157.77*
Front fog lights	*£96.06*
Inertia reel rear seat belts	*£78.10*
Nylon rug to the rear	*£117.10*
Reading lamps (pair)	*£43.61*
Tinted glass all round	*£299.00*
Leather trim to the rear	*£436.05*
Cloth trim to the rear	*£249.17*

Also available (but not listed) were a blind to the division, cocktail cabinet, fire extinguisher, flagmasts (with three optional mounting points), intercom telephone, and a pair of writing tables. A PVC washable headlining could also be specified at no extra charge.

The proposals from Keith Cambage were carried through, and in March 1982, a new price list was produced to correspond with the reports findings. This was the start of better times for the limousine. Cambage, who had started at Jaguar as an apprentice in 1951 (and for a time worked with the racing team in that legendary era of the C and D types), had literally saved the DS420, because as a result of these proposals and the constant updating of the limousine, sales once again picked up to an acceptable level.

The Office Car

In January 1984 the Daimler limousine *"office car"* was first seen. It was built in association with David Earl Associates of Wellingborough and STC Electronic Services of Harlow, who managed the electrical aspect. A top-specification DS420 was used, with individual seats to the front, and two vast armchair-style seats to the rear, both trimmed in biscuit leather. The crystal glassware in the rear centre console took second place in the priority listings, with the two compartments between them sporting a colour monitor, video player, two radio cassettes and an amplifier, two computers and a printer, and a Pye radio-telephone. The car was built for publicity and indeed, once this purpose was served, the car was gutted and rebuilt as a standard vehicle. Complete, the office car's price was about £49,000.

Better Times

Things were indeed looking up for the Limousine Division; a total of 169 cars were built in 1984, compared with 168 and 119 for 1983 and 1982 respectively. This marked quite an achievement considering that so much effort was being poured into the XJ6 replacement, the XJ40 project. The actual delivery figures for 1984 were as follows; 135 limousines worldwide, with 123 of them staying at home - 34 drive-away chassis also found new owners.

The 1987 Model Year vehicles, saw several major specification changes. Firstly, a cathodic undercoat was introduced to prolong the life of the coachwork, and seam sealing was also put into use with the same aims. At first the cathodic dip caused a few problems; the treated body had to be baked to at least 185 degrees C, but the lead used melted at a lower temperature. Several different leads were used but none were found to be suitable. Park Sheet Metal, however, found that a sheet of metal treated on both sides, could be lead-

Below:
Keith Cambage - Director of Limousine Operations.

Right:
One of the face-lifted 1980 model year cars.

Left:
A rear view of the above car. By now, all production had been moved up to the Jaguar works in Coventry.

loaded on one side without affecting the other. Most rust problems actually start on the inside of a motor car, so the DS420 was dipped and then leaded from the outside in the knowledge that the inside of the panels were 100% protected - Jaguar then treated any outside areas as necessary. As a result, a six year warranty on bodywork became a standard feature of the car.

There was also a major revision of the lights and bumpers in 1987, both front and rear. The bumpers were now similar in design to those used on the Series 3 XJ saloons; the grille, too, was remodelled to look more modern. The heavy duty Bosch alternator was by now used across the board, when previously it had only been used when a vehicle was fitted with air-conditioning.

Above: *A body nearing completion at Park Sheet Metal works, situated between Coventry and Bedworth.*

Warning lights were revised, and so was the locking system with the introduction of two key locking - previously five or six keys was the norm. Also the Daimler badge was now gracing the boot to the right of the number plate.

Fewer limousines and hearse chassis were produced during 1986 - a total of nearly 175 vehicles - as a great emphasis was being placed on the introduction of the new Jaguar XJ6 model. However, a number of significant units were built, the most important being a replacement vehicle for HM the Queen Mother. A limousine was also built for Queen Alexandra of Denmark, finished in Windsor blue, and Jaguar's French distributor had a divisionless model made for a town car. Sir John Egan's limousine was completed in May 1987, a month after the price of a standard DS420 had risen to £33,100. This was a special divisionless car, absolutely packed with extras. The rear fin fitted on the boot lid was in fact an aerial for all the equipment inside.

By now, the somewhat staid colour schemes of 1968 were a thing of the past. Standard coachwork colours still included black and Carlton grey, but cirrus grey, regent grey metallic, grosvenor brown, antelope metallic, Jaguar racing green, Windsor blue and steel metallic, were all added to the range. Leather trim could be specified in either barley, biscuit, black, buckskin, doeskin, Isis blue, magnolia, mulberry or saville grey, with cloth trim in West of England fawn or grey, and dralon in beige, dark blue, light blue or maroon.

At the start of 1988, Wilcox Limousines joined with Eagle Specialist Vehicles of Bolton to create a hearse about £4,000 less expensive than the competition. This was done by building the roof and side areas in glassfibre - other specialists followed suit shortly after. The 1988 Motor Show (held at the NEC, Birmingham) saw the 1989 Model Year Daimler limousine win the Gold Medal for coachwork, just as it had in 1986. Finished in two-tone blue, it was one of five vehicles to win honours on the Jaguar stand.

Several modifications were introduced for this particular model. The Jaguar five-spoked alloy wheels (adapted to carry nave plates) became a standard feature, replacing the old pressed steel type. This helped make the car look more modern and attractive, and had the effect of reducing costs through standardisation. Attention was focused mainly on the fascia details; the choke was relocated for ease of operation, a new warning light cluster was introduced, and the hazard lights switch was moved onto the new modified Series 3 XJ steering column. Options were still very much the same, the only notable addition being piping for the leather trim. The total price for the standard limousine of 1988 stood at exactly £40,000.

Very little changed in the production of the DS420 limousine - it was almost completely hand-built. After body parts were received from Motor Panels, it took nearly three solid working weeks and many skilled craftsmen for the car even to resemble a finished vehicle.

The structure was built at Park Sheet Metal using original jigs, it was then joint welded, planished, filed by hand and finished by eye. Once the panels had been hung, the car was transported from PSM to Motor Panels, where it received its cathodic dip. When the car returned to Park Sheet, the lead loading was carried out and tolerances met. The car was then cleaned and inspected, ready for shipment to Jaguar.

The body-in-white, on arriving at Browns Lane, was inspected once more and adapted to suit the customers needs (ie, left or right hand drive, air conditioning etc.). It was then washed thoroughly and force dried. Anti-corrosion measures were applied; the body was dipped up to waist level, stone-chip paint sprayed onto vulnerable areas; wax and underseal protected all the joints and box sections, and polyester was sprayed onto any lead loaded parts of the body to prevent paint reaction. After the vehicle had been sprayed with primer and flatted down, the colour coats were then applied, followed by two coats of clear lacquer to give depth of shine.

Electricians made and installed a wiring harness

Right:
Only the very best craftsmen are employed to work on the Daimler limousine. This beautiful cabinet, for example, was built from scratch to a one off order.

Right:
With the North American market in mind, this prototype was built late in 1984. Had it reached production, twin bumper blades with rubber inserts would have been used.

for the car as each one was unique. The fuel tank, petrol lines and brake pipes were then located before the suspension and power-train could be fitted. All the glass was then put into place, and following a successful water test, trim was inserted, any leather being cut and sewn by hand.

Final fitting of the panels was followed by a thorough road-test, every few cars being tested by Keith Cambage himself (up until his well-deserved retirement in February 1991); while the engine was still warm, emission tests were carried out. Two men then fitted the bumpers and grille, and the car was inspected once more before its delivery. By these methods, the Daimler limousine was built to very high standards. When the build was at Vanden Plas, the internal Jaguar audit rating was very low, but in the 1980s it was well above any other model that the Company produced.

Royal Cars

The list of personalities who have used the Daimler limousine is indeed impressive, and includes Royalty, Heads of State, Prime Ministers, Governors, Ambassadors and diplomats to name but a few. British Royalty have once again started to support the marque which is very exciting. During the 1970s, the Queen Mother was the only member of the Royal family using Daimler limousines, which were all supplied through Coombs of Guildford - famous for their tuning conversions of Jaguar Mk.IIs.

The first Royal DS420 was chassis no. 1M 1559 with body no. 4M-1496. Finished in Royal black over Royal claret, it was delivered to HM the Queen Mother in 1970. Distinguishing features included the coat of arms plaque mounted over the discreet blue police light, and the Queen Mother's lion mascot in place of the flying D. A red coachline replaced the chrome lower waist strip, and drip rails were finished in black, again replacing chromework. Side repeater flashers were also used, mounted on the front wings. Inside, dark blue cloth trimmed the front compartment, while to the rear, fawn cloth was used. Many of the standard options were chosen, including the use of electric windows all-round (including the glass division). The modification of the rear centre armrest was necessary to incorporate a Plessey radio and vanity mirror. Specially-designed electrically-controlled steps were built and fitted to the car - the original design work coming from Coombs, with Jaguar refining the system. Eight years, and over 70,000 miles later, the vehicle was replaced by body no. 4M-3351. It retained the NLT 2 registration number, along with all the special fittings of the previous vehicle, many of which are fitted to the present car.

In September 1983, HM the Queen Mother took delivery of the first Jaguar-supplied vehicle; the body number was 4M 001468, and the chassis SADDWATL3AC 200453. This was to be the first of five Daimler DS420 limousines delivered to the Royal family in the 1980s.

Another car (also NLT 2) was delivered in December 1986. Air-conditioning was added to the features present on the Queen Mother's original DS420, and Royal burgundy was the specified colour for the lower half of the coachwork. Chassis no. 200905 kept the tradition of Royal Daimlers very much alive.

HM the Queen had three limousines herself over the last decade. The first one (chassis no. 200477, body no. 4M 001479) was delivered mid-way through 1984, and was finished in the traditional Royal colours of black and claret. It was finally withdrawn from service three years later, when D500 GUW rolled into the Mews (chassis no. 200970).

The third Royal limousine supplied to HM the Queen during this period, was delivered in August

Above:
Chassis no. 201127 was delivered to HM the Queen in 1988. Another DS420 was delivered shortly afterwards.

Left:
The face-lift in 1988 gave the DS420 a more modern look. Externally at least, it would remain virtually unchanged for the next few years, becoming the final form of the type.

1988. Chassis no. 201127 had the body no. 4M 002128, and carried the registration F400 OYO. As one would expect, the vehicle was full of optional extras, including air-conditioning to the front and rear compartments, electric tinted windows and division, an electronic audio system, central door locking, reading lights, lambswool rug, and of course, the blue police light, flag mast and illuminated plaque on the roof - a feature on all Royal cars. The coachwork was finished in black and Royal claret, and in the Queen's cars, the front trim was in dark blue cloth as usual, but the rear compartment was specially finished in Baroda blue.

The End of an Era

It became noticeable that the Daimler limousine was being used increasingly for official duties, not only by members of the Royal Family but also by visiting dignitaries - one could be seen almost every week on television. The days of the models future, however, had been numbered for a great many years. The DS420 was the last non-military user of the fabulous XK engine, but that alone could not save it, nor did the fact that it was the only real hand-built limousine in the world - other vehicles being stretched saloons in effect.

The XK engine incidentally, remained in use for the limousine, because the V12 engine would not fit, and the 3.6litre AJ6 unit couldn't offer the necessary low-down torque. The 4litre version of the AJ6 would, but unfortunately, the DS420 was losing pace with modern safety laws, and proved too expensive to update.

Several years ago, the author was told that a stretched XJ40 had graced the drawing boards, though it was very doubtful whether this would be built at the factory, it being far cheaper and easier to let an outside firm foot the development and tooling costs. Sure enough, in the early part of 1994, Eagle Specialist Vehicles produced a limousine on the lengthened 4litre XJ40, about six months after the last DS420 had been built. A hearse on the same chassis was also available for £2,000 more than the limousine, at £57,000.

The last DS420, chassis number DWATL3AA 201631, which left the Browns Lane workshop in November 1992, is being kept by the Jaguar-Daimler Heritage Trust, while the two before it went to the Royal family; one to HM the Queen Mother, and the other to HM Queen Elizabeth II. What, I wonder, will the Royal Family order in the future?

There seem to be no plans in the pipeline for a direct replacement, but with the new management, this may still change. Jaguar still have the Lanchester name at their disposal of course, and it will be interesting to see if that is revived, too. The first *Motor* magazine article on the DS420 noted that it should last as Daimler's limousine model for at least the next two decades. At the time this must have seemed a little far-fetched, but looking back, how right it was!

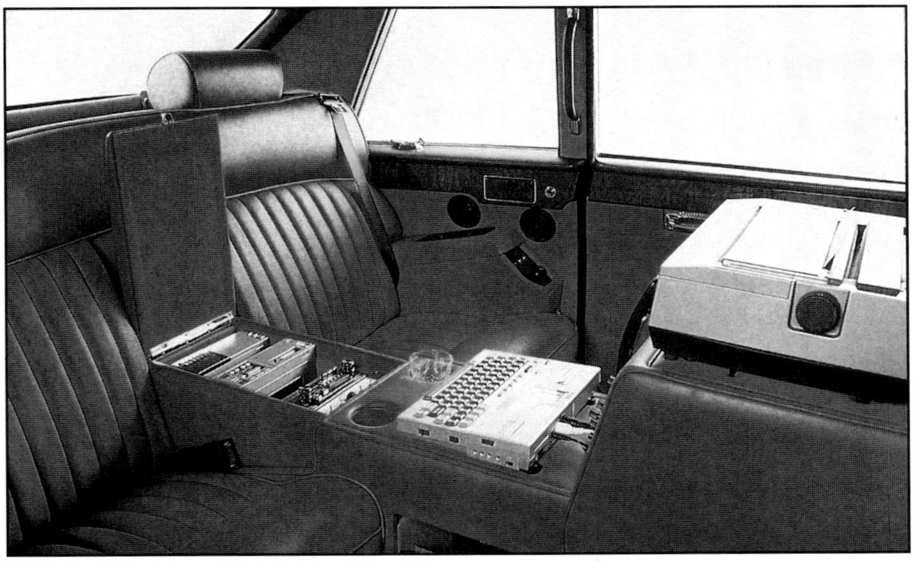

Left:
Dubbed the "Office Car", this must surely rate as being the world's ultimate executive toy.

Below:
The last cars to be delivered to clients the Daimler marque; on the left, a DS420 for HM the Queen, and to the right, a similar car for HM the Queen Mother. The final car to be produced has been retained by the factory.

Chapter Fourteen

GRACE, SPACE AND PACE AT ITS BEST

Billed as *"the Sovereign for the Seventies"*, the XJ6 based Daimler was introduced a year after its Jaguar brother in 1969. Design work on the XJ project began in 1964; Sir William Lyons was the designer and Bill Heynes the engineer - it was to become the highlight of both their careers. Introduced just four years after the project had begun, the new Jaguar XJ6 was to take the motoring world by storm. On 26th September 1968, Sir William unveiled his new creation. Each successive night, the Royal Lancaster Hotel in London, was filled with the rapturous applause of the guests, as one by one, XJ6s were picked out by spotlight, until finally all ten vehicles were on show.

Dunlop were asked to design special radial tyres to suit the vehicle's characteristics. Their new ER70 VR15 tyres gave much better roadholding and handling, and after Bob Knight managed to eliminate the problems associated with radial tyres at the time, they put Jaguar at least four years ahead of Rolls-Royce, who were still using crossply rubber.

The XJ6 was voted Car of the Year, and also gained the Don safety award, which didn't please Volvo too much! More praise was heaped on the car when, on 9th October 1969, a Daimler version was introduced, the Daimler Sovereign.

Facing Page: *A Jaguar works publicity photograph of the XJ40 Daimler 4.0.*

H940 DKV

The body was built at Castle Bromwich by Pressed Steel Fisher, with the mechanical specification owing much to its predecessors. The Mk.X became a mobile testbed, the front suspension coming from that vehicle, the only difference being that the shock-absorbers were mounted outside the coil springs. This had the effect of reducing the inclination to dive under braking by about 50%, and because of this, softer springs could be employed thus improving ride quality still further.

Engines of 4.2litre and 2.8litres were selected. The 4.2litre XK unit was a slightly modified version of that fitted to the Jaguar 420 and Daimler Sovereign, with most of the changes being limited to the cooling system. It is because of this engine that the now-familiar power bulge appeared - on the first occasion a long-stroke 4.2litre engine was fitted to an XJ6, it was found that the bonnet line was too low (a new engine had been planned), so the elegant bulge was put in the panel to solve the problem. The smaller capacity of 2.8litres was chosen to avoid the tax laws in certain European countries. With the exception of North America, who used Zenith Stromberg 175CDs, twin SU HD8 carburettors were employed to supply the engine with fuel.

At the launch of the XJ6, it was stated that in two years time, new additional power units would be made available; several options were tested, including V8 and V12 engines. A slant six, based on half of the V12 was also looked at, and the engine bay was made large enough right from the start to accept any one of these options. Strangely, the very first XJ, chassis no. P1G 1001 DN, is listed as being a Daimler fitted with a V8 engine. Finished in silver-grey with a black interior, the vehicle was passed to the Experimental Department in June 1968. The car was never registered, however, and eventually had a 2.8litre XK unit fitted. The V8 inherently lacked the balance needed for a Jaguar engine and besides, competing head-on with the Americans seemed pretty pointless. Only the development of the V12 would continue.

The New Daimler Saloon

The general outline of the new Sovereign was identical to that of the Jaguar, but appearance was altered by an attractive version of the traditional Daimler fluted grille, with the fluted theme being carried over to the rear boot-lift. The nave plates carried the famous D insignia, and the different badging to the rear of the car also helped to identify the new model. Again, the 4.2litre (245bhp) and 2.8litre (180bhp) engines were offered, and if a manual gearbox was specified, overdrive was fitted as standard.

The interior also differed slightly from that of the Jaguar. Detailing features included wider pleats in the Slumberland-designed leather seating, larger armrests, leathercloth door cappings, and a carpeted kick-pad on the doors. By August 1969, a provision had been made for the optional headrests on the XJ6, although some of the earliest Sovereigns didn't carry this modification. There was generally a higher quality finish on the Daimler models, but traditional Jaguar values were applied to the new model - those of outstanding performance, combined with quality and value for money.

Right:
One of the many early styling mock-ups. Close inspection reveals several features which would later be carried over into production models.

Left:
Testing the new car on the deserted roads of Wales.

Below:
A new Daimler for 1969 - the Sovereign.

RDU 5H seems to have been the car used for the first publicity shots. The American market was never exploited with the Daimler - the expense and complications of qualifying the Sovereign to be sold on that market were not considered to be worthwhile. As a letter was posted from the factory, advising dealers of the new car's prices and options, the yearly sales figures were released and somehow, a few XJ type Sovereigns appeared on the ledgers - all six of them went to distributors in Italy. On the Daimler stand at Earls Court, a Sovereign was displayed alongside a DS420, and another fine example finished in Regency red with a beige interior, was shown on the Pressed Steel Fisher stand.

Before the XJ6 was introduced, Jaguar had employed the policy of demand through scarcity; they would offer many different models and variations. In 1970, however, the 420G was discontinued,and the only saloon models to continue in production were those based on the XJ.

The first delivery figures were very encouraging. At home, 928 2.8 litre cars were sold, and 1,450 4.2 litre cars. On the export front, the totals came to 151 and 166 for the 2.8 litre and 4.2 litre models respectively. The comparative rarity of the Daimler model becomes obvious when one thinks that for every Daimler built, nearly five Jaguar versions were produced, and by the 1980s, the ratio had reached six to one.

Production Changes

Many detail changes had already been applied to the Sovereigns. Some of the most important modifications included the fitting of anti-surge flaps to the petrol filler caps following complaints about the smell of petrol fumes; the front wheel arches were slightly modified to give the large tyres a little extra clearance, and ventilation was improved by the adoption of fresh air vents to the footwells, the intakes being cleverly disguised at the top of the outer headlights. Tread plates were fitted, as was a new three-piece rear bumper - the joins being covered by the overriders. HS8 carburettors replaced the old HD8s; a new automatic enrichment device (aed), and quieter running camshafts were employed. In addition, the Model 12 automatic gearbox superseded the Model 8 which had been used originally.

Left:
This superb Double-Six was displayed at the launch of the Jaguar XJ12, held at the Skyline Hotel. The Double-Six was officially announced one month after the XJ12.

Below:
A Series 1 Daimler Sovereign in an evocative period shot.

The familiar pairing of the Sovereign and DS420, was accompanied at Earls Court in 1971 by Lord Montagu of Beaulieu's famous 1899 Daimler 12hp model, put there to celebrate the Diamond Jubilee of the Daimler Motor Company; 1896-1971.

During April 1972, *Autocar* published a full road test on a 4.2 litre Sovereign with automatic transmission. The report was very complimentary, the last paragraph stating: *"In our original test of the Jaguar XJ6, we concluded that the car represented a tremendous advance, putting it ahead for several years at least. As tried again in its Daimler guise, we find it still well in the lead for its unrivalled combination of virtues."*

The V12 Reborn

The Jaguar XJI2 was introduced in July 1972. This vehicle was fitted with the 5.3litre, high performance V12 engine, first seen in the E-type sports car just over a year earlier. Together, Bill Heynes and Walter Hassan, designed, developed and later productionized this magnificent engine. Four- and two-camshaft engines were compared in the Mk.X; the four cam version being essentially two XK units put together. However, after the XJ13 racing programme had been brought to a halt, it was decided that the two camshaft engine (one per bank) was the one to continue with. The two-cam flat head design was not only found to perform just as well as the other engine up to around 5,000rpm, but was also cheaper to machine, less complicated to build and service, quieter and lighter. It also developed a higher torque output at lower revs, which was an important factor to consider, as an automatic gearbox was to be employed. A manual gearbox had been designed, but at the time, there was no overdrive unit available which could possibly handle the massive torque.

It had been hoped to use fuel injection from the start. The Lucas system gave good performance, but apart from the fact that it would never have passed the new US Federal legislation, the metering system was not sufficiently accurate. Brico came up with the perfect solution, but cancelled the firm's facility at the last minute. Four carburettors were therefore employed just to get thing rolling.

As the V12 engine was at least 80lb heavier than an XK unit, the spring ratings were adjusted accordingly, and higher rated tyres and better front brakes improved safety, as this was to be a very fast car - in fact, the XJI2 was the fastest production saloon in the world. It came as no great surprise that the XJI2 was voted the Car of the Year for 1972.

One month after the XJ12 launch, a Daimler version was announced, and its revived name (chosen by Lofty England) was the Double-Six. Introduced at the very low price of £3,848.90, the Double-Six had nevertheless become the highest-priced XJ to date. In automatic form only, for the reasons described earlier, the Double-Six 5.3litre V12 engine developed an incredible 309bhp at the flywheel. Front seat belts were fitted as standard, and air conditioning was listed as an option. The prices were revised on the smaller models - the 2.8litre model with manual/overdrive transmission was put at £2,953, and the asking price for an automatic version was £3,009. An automatic 4.2litre Sovereign would have cost £3,358, while with manual/overdrive,

Below:
The long wheelbase version of the Sovereign was to prove very popular. Introduced in October 1972, it formed the basis for many of the XJ models that followed.

Above:
The ultimate XJ so far in the story - the Double-Six VDP. Some 337 of them were built for the home market between September 1972 and September 1973, with a further six going abroad.

the price was reduced to £3,259.

The official 50th anniversary of Jaguar was due on the 4th September 1972, but within a month of the celebrations, due to the corporate policies of British Leyland, Jaguar Cars Ltd ceased to be a separate entity, although it did at least retain its identity. Lofty England took over from Sir William Lyons, firstly as the Managing Director, and later as Chairman. Later, Lofty England also retired when Geoffrey Robinson was called in to run the Company.

The Double-Six VDP

On 26th September 1972, another new Daimler model was officially introduced, although the Press had actually had a sneak preview at the XJ12 announcement. At last, the ultimate luxury vehicle had arrived in the form of the long-wheelbase Double-Six Vanden Plas. The VDP model had an extra 4 inches inserted into the wheelbase, with the benefits going to the rear compartment.

Mechanically, it was similar to the basic Double-Six , but the trim was pure luxury, with many extras

Left:
Putting an interesting new model through its paces on the picturesque Swiss highways. Note the Double-Six badge to the right of the number plate.

Left:
A prototype two door coupé. There are drawings at the Jaguar works of a Series 1 two door model, but due to time running out on the old design, only the Series 2 version made it into production.

Below:
The interior of the Series 1 Double-Six Vanden Plas. Note the extra woodwork in the door cappings and the 160 mph speedometer found on the V12s.

being offered as standard, including air-conditioning, four speaker radio/cassette, electric aerial and windows, heated rear backlight, laminated windscreen, Sundym glass, remote control door mirror, fog and spot lights, headrests, rear reading lamps, quartz halogen headlamps, vinyl roof, and front seat belts. Chrome-plated wheels and whitewall tyres were also offered.

To make them even more exclusive, the Double-Sixes left Jaguar in an untrimmed and unpainted state, and were transported to Vanden Plas. Once at the Kingsbury Works, they would receive their special trim, a vinyl roof, and one of seven unique coachwork colours - three top coats were applied. Silver sand, caramel, coral, Aegean blue, sage, morello or aubergine could be chosen, although the vinyl roofs were always finished in black. The colour of the leather trim was also to the customers choice, with the options being chamois, tuscan or deep olive. The Vanden Plas model was rather expensive at £5,439.06 when compared with an XJ12 at £3,725, but in the company of Rolls-Royce, Mercedes-Benz and the like, it was still very reasonable.

More New Models

By 1972, the 4.2litre Sovereign was becoming very popular (well over 3,000 had been sold during the 1971/72 season). Following the launch of the new range of Mercedes-Benz long wheelbase models, however, Jaguar decided to follow suit, and introduced a new long wheelbase version of the Sovereign. The 4.2litre Sovereign LWB was announced in October 1972, and the same 4inch extension used on the Double-Six VDP was chosen in the interests of standardization. Priced at just £3,585 for the manual/ overdrive model, and £3,684 for the automatic, these cars were soon to prove good sellers.

During April 1973, *Motor* magazine ran a test on the Double-Six. They praised the car's exceptionally high performance, superb quietness and refinement, road manners and comfort, but were a little put off by the rather muddled minor controls, an *"indifferent"* heating system, and the restricted rear leg room. Later on in the month though, the same magazine compared a Vanden Plas model with a Rolls-Royce Silver Shadow, and in all, the cars came out fairly evenly matched. The Daimler was given the verdict as being the better driver's car, while the Shadow was noted as being slightly better for the rear passengers. Summing up, it was noted that it was *"not an easy choice."*

By August 1973, the 2.8litre Sovereign had been

dropped from the range, although it still appeared on price lists for the benefit of dealers who still had one in their showroom. The 2.8litre Sovereign had been dogged by a piston holing problem caused by residual metallic deposits, which resulted in pre-ignition. Timing and piston modifications did cure the trouble, but the bad publicity had affected the models popularity. The last V12 SWB saloon was produced (as a Series 1) during the following month, when the dealers were notified of a new and exciting line-up of Daimler cars.

The Series 2

New American crash regulations led to the introduction of the Series 2 models. Unlike the Series 1, which was kept a complete secret until the actual launch, the new Series 2 had a Press preview held in Aberdeenshire. The first public appearance, however, was at the Frankfurt Motor Show on 13th September 1973.

One of the new US Federal laws stated that bumpers had to be a standard height of 16inch from the ground - Jaguar achieved this admirably (which is

Left:
A rather rare Series 2 SWB Sovereign, built for just one year after being introduced in September 1973.

Right:
The elegant, and now very collectable, Double-Six two door model. Although this example dates from 1974, production did not begin until early 1975 and by this time, the V12 coupés featured a chrome moulding along the waistline.

Right:
Now the property of the Jaguar-Daimler Heritage Trust, this superb early Series 2 Double-Six Vanden Plas was once owned by the late Sir William Lyons.

more than can be said for some of their competitors) by using a shallower grille, and fitting the overriders under the bumper - this arrangement was then tidied up by an undergrille. The rear number plate light also had to be repositioned so that it was incorporated into the actual boot-lift. New and stricter anti-pollution measures had to be taken; air had to be heated before induction and controlled thermostatically, which unfortunately led to reduced power on all engines. Although these features aimed at the American market were of no real importance to the Daimler marque, it not being sold in the USA, the cars had to be updated in the interests of standardizing the range.

Much thought had been given to the safety features of the Series 2 cars; doors now gave extra protection against side impact through the use of welded in W section barriers, and a laminated windscreen was now fitted as standard throughout the range. The braking system was totally upgraded, with ventilated discs being fitted on the front wheels, thus improving the car's stopping efficiency. Heavy padding was applied to all areas of the interior where injury might be caused in an accident, and noise was cut down even further by the use of specially moulded fire resistant coverings to the floor and bulkhead.

A new introduction was the use of fibre optic cables, seen for the first time on a British vehicle, which illuminated certain control areas of the new modernized dashboard. Two column stalks now replaced the classical bank of switches associated with the Company for so long - the Press were happier, but some purists were understandably upset.

A completely new and interesting coupé was also shown at Frankfurt in September 1973. The two-door XJ model was based on the SWB floorpan, and was to be Sir William Lyons' last styling project. The car had been slow to reach production because the wide doors and complex window winding and sealing (Sir William insisted that the car should remain pillarless), took a lot of development. The coupé eventually went into production about a year after its debut.

At Earls Court, a good cross-section of the Daimler range was on show in 1973. The Sovereigns were represented by a standard saloon, a LWB saloon, and a two-door model. A Double-Six coupé was also exhibited, along with its LWB and VDP counterparts, and of course, a DS420 limousine.

A price list was released in December 1973, although prices continued to fluctuate slightly over the following few months. By the time the first real Series 2 delivery figures were released, reflecting the period from 1st October 1973 to 30th September 1974 inclusive, the SWB Sovereign had been dropped from the range.

Left:
This is the Lynx Convertible based on the two door XJ - the Avon version was very similar.

Right:
The production line at the Vanden Plas works in Kingsbury, pictured here during the mid-1970s.

Left:
A splendid example of a 1976 Sovereign. It should be noted that the 3.4 litre Sovereigns were identical to the 4.2 litre models in their outward appearance.

Left:
The Series 3 XJ saloon was introduced in the early part of 1979. This is one of the very first Daimler versions, seen here in the Welsh countryside.

Right:
The V12 HE (High Eficiency) engine.

Below:
A Double-Six HE from 1983.

The figures read as follows:

		U.K	Export
4.2 Sovereign SWB	*(O/D.):*	*85*	*38*
	(Auto.):	*1,811*	*233*
4.2 Sovereign LWB	*(O/D.):*	*33*	*9*
	(Auto.):	*1,786*	*287*
Double-Six	*(Auto.):*	*758*	*210*
Double-Six VDP	*(Auto.):*	*531*	*61*

From these figures it would appear that the 4.2 SWB model was the best selling, but it was dropped because it was said to be unpopular. Standardization is perhaps the more likely reason, despite the fact that the two door coupé, which used the SWB floorpan, was nearing production.

During October 1974, *Autocar* magazine published the first full road test on a Series 2 model. The Sovereign they chose was fitted with the Model 65 automatic transmission, which by now had replaced the old Model 12 unit. Summing up their report, they stated that the Sovereign was *"a most desirable car in any company."*

Changes in the Range
There was a flurry of new models in April 1975. The two-door coupé at last made its official launch, although its production was to be limited because the SWB floorpan was no longer readily available. The Double-Six version was to sport fuel-injection; the system used originated from Bendix - it was then developed and engineered by Bosch, before Lucas and Jaguar finally adopted it. Fuel injection restored the V12s power from 250bhp to about 285bhp. The saloons, however, retained their carburettors for a while longer. The suspension was very slightly uprated, and the rear axle ratio was changed on the fuel-injection cars from 3.3:1 to 3.07:1. A black vinyl roof covering and chrome coachline were standard fitments, and a discreet fuel injection badge to the rear of the car, immediately signified what was under the bonnet.

Due to yet another oil crisis (this time an Arab

oil price increase caused panic in the UK), a 3.4litre model was introduced as an alternative to the larger 4.2litre and 5.3litre cars. With the model retaining the Sovereign name, the 3.4litre unit developed a healthy 161bhp. Although this was in effect an old Jaguar engine, the B.4 unit had a totally new block, made with the same XK tooling as the 4.2litre. It was offered with either a manual/overdrive gearbox, or Borg Warner Model 65 automatic transmission. Cloth trim, available in five colours, was fitted as standard. The vehicles top speed was approximately 115mph, and an average of 20mpg was not out of the question. The speed from 0-60mph was similar to that of the 4.2litre Sovereign, but once cruising there was a noticeable difference in top gear acceleration.

The fuel crisis led to the assumption that the V12 engine would have to be gradually phased out by the early 1980s. With this in mind, Jaguar introduced yet another new model - the 4.2litre Vanden Plas. Finished to exactly the same standard as its Double-Six brother, the 4.2litre Vanden Plas offered the best in luxury, combined with a notable improvement in economy. The price on introduction was £7,130, about £900 cheaper than the Double-Six VDP.

Despite rising prices, sales increased. The 4.2litre Sovereign was always the more popular model, usually selling something like 2,500 units in any given year, which was nearly the same total as all the other models put together. The GM400 automatic gearbox was introduced for the V12 cars during March 1977, and this gearbox, which is recognized as being one of the best in the world, was also used on the Daimler limousine.

By now, Jaguars and Daimlers were being exhibited at Motor Shows alongside other lesser Leyland offerings, and prices were still climbing at an alarming rate. However, the news that the Double-Six Vanden

Above: *This picture shows the revised control panel of the 1983 cars. This basic layout was retained until the end of the Series 3 run.*

Below: *A Series 3 Sovereign dating from 1980.*

Above:
The Daimler S, or XJ-S with a difference! This model came about when Jaguar were looking at diferent ways of marketing the XJ-S. The cabriolet and convertible evolved from this research, and went on to be highly successful. Unfortunately for the Daimler version, public reaction was tested in the USA, and because Daimler had no market there it was immediately placed at a disadvantage. Only one was ever built.

Right: *The Daimler 4.2.*

Plas had been voted the Best Car in the World by *Car* magazine, must have helped morale to a certain extent.

Due to more Federal laws, 4.2litre cars shipped to the USA after May 1978, had to be fitted with fuel injection. This not only reduced the emission levels, but in making the engine more efficient, also made a noticeable improvement in fuel consumption, too. Fuel-injection was not introduced on the home market, however, until the Series 3. By the end of the year, the 3.4litre Sovereign, and both the coupé models had long since disappeared from the price lists - the 3.4litre engine was limited to the Jaguar range, while the beautiful two-door models had gone forever, simply because they had become uneconomical to produce in such small numbers.

The prices of the Series 2 Daimler saloons rose for a last time on 11th December 1978. The prices ranged from £11,646 for the Sovereign (with either an automatic or Rover SD1-type five speed gearbox), right up to £18,219 for the flagship model, the Double-Six Vanden Plas.

At least one Series 2 Daimler is assured of survival - 2S 11348 has been kept in the Jaguar Collection as an example of a modern Daimler. It is interesting to note that a lhd two door Double-Six was once presented to the Board in full VDP trim for approval. Unfortunately, the idea was rejected and the car disposed of.

Avon (one of Sir William Lyons' chief rivals in the early days) revived the coupé for a while. The Avon Body Company of Warwick was formed in the early years of the 1920s, achieved fame in the 1930s, struggled through the war years, but carried on right up until 1973, when Graham Hudson took over and incorporated it into a new company called Ladbroke Avon. The Special Projects Division was to involve itself with such challenges as creating a soft top version of the XJ-type two-door coupé.

This convertible model was produced between 1979 and 1986, with the full backing of Jaguar. The conversion cost £6,500, and only 25 were built, ten of which were Daimlers, with one of those being a rare lhd model. The cars were painted in the colour of the customers choosing, and the manually operated hood was made from PVC and mohair. Another company called Lynx Engineering, famous for their C and D type replicas, went on to do much the same thing.

The Series 3

'The Series 3 is a beautiful synthesis of everything that Daimler was, and everything that today's owner and driver will expect his Daimler to be.'

That was how the first catalogue for the new cars put it, but in reality, there should never have been a Series 3 at all. The Series 3 was introduced in the interests of self preservation - the next car should have been the XJ40 (or new XJ6) but because Jaguar knew that Leyland would have failed to do the car justice, the project was stalled.

Launched in Torquay towards the end of March 1979, the cars styling had been put out to the Pininfarina studio, perhaps more famous for their Ferrari designs, the subtle changes the Italian studio made to the XJ were certainly no less monumental than some of their finest masterpieces. Although the cars looked very much the same in profile to the LWB saloons of the previous series, few of the panels were actually interchangeable. New bumpers with black inserts (similar

Left:
An AJ6 engine in 3.6 litre form, as first seen in the Jaguar XJ-S sports car before being adopted for the new saloon.

Below:
Project XJ40 at a very early stage. This is a Giugiaro design proposal dating from June 1974, and looks very much like a Maserati. Giorgetto Giugiaro himself admires Jaguar styling, and said in a recent interview that it was his favorite car.

to those of the XJ-S sports coupé) with the indicators recessed in the front one, larger rear lamps, and flush fitting door handles were instant distinguishing features. A flatter roof and a more elegant rear window line were also noticeable. A massive improvement was gained at the front by the neat treatment applied to the area underneath the bumper - gone now was the undergrille that the motoring writers loved to hate; it was replaced by a thin chrome line. The front door window was now in one piece, with all glass being tinted as standard, at least on the Daimlers.

Complying with traditional Jaguar practice, the new cars were mechanically much the same as their predecessors. The main difference was the standardization of Lucas L-jetronic fuèl-injection across the range. Applied to the 4.2litre XK engine, this would allow a higher compression ratio and larger inlet valves to be fitted, thus increasing the power output by around 25bhp, while at the same time, making the car more economical.

Inside the Series 3, map pockets were installed in the backs of the front seats (which were now electrically-adjustable for height on the VDP models), and there were wider pleats in the leather trim. The door mirrors had remote control, and four door-mounted speakers were now fitted as standard equipment. The controls on the column stalks now complied with ISO guidelines, and incorporated a six second intermittent windscreen wiper speed for the first time; something the Series 2 had needed badly. The central locking was now by means of a key, and on entering or leaving the car, a courtesy light delay switch automatically lit up the interior for a few seconds.

The bodywork on the VDP models was fully lead-loaded and carried an even more luxurious finish than before. 'TPA' paint was being used on all Series 3 cars, and the new and attractive standard coachwork colours included: Tudor white, Cotswold yellow, damson red, racing green metallic, cobalt blue metallic, quartz blue metallic, chestnut brown metallic, and rhodium silver metallic. As always, the Vanden Plas saloons had their own unique coachwork colour range, with silver sand, caramel, amethyst, biascan blue, mistletoe, mink, and coral being offered. These colours would stay much the same for well over two years.

As expected, prices were soon revised; a hefty increase of just under 12% took effect in June. Even after this, however, the prices were still very reasonable compared to the nearest competition, notably from Mercedes-Benz. The most important change made at this time was to the automatic gearbox, which was replaced by the Model 66 for the six cylinder cars. The following September, *Motor* magazine published a full road test on an automatic 4.2litre Sovereign. The magazine reported favourably on the car, and concluded;

Left:
The new model that Jaguar connoisseurs had been waiting patiently for - the XJ40. This is the flagship version, the Daimler 3.6.

Right:
The sumptuous interior of the Daimler 3.6, the ultimate in sheer luxury.

"This in many ways is the best saloon car in the world, and most certainly the best value for money in its price range."

Leyland had very nearly destroyed Jaguar and Daimler - even the XJ-S launch in 1975 was a distinctly low key affair. Under the leadership of Sir Michael Edwardes, however, things began to improve slightly and, by 1978, the name of Jaguar was again being recognized by the BL hierarchy. A plan was then put forward to combine the specialist manufacturers, ie. Jaguar, Rover and Triumph. The proposal was approved, but the arrangement was short-lived. By now, the morale of the workforce was very low, but the appointment of John Egan early in 1980, soon began to make a welcome difference.

Special XJs

At the 1980 Motor Show in Birmingham, the Avon Body Works gained a Gold Medal in the International Coachwork Competition with an XJ Estate, designed by Tony Stephens (who had earlier penned the Avon XJ Convertible) and based on the Series 3 saloon. The Avon Jaguar-Daimler Estates cost £7,500 to convert, and each one was built to the customers own specrlication, with Avon themselves supplying the original saloon. Production of this vehicle was limited to a run of 250 models of both Jaguar and Daimler combined.

The Series 3 XJ was converted into an estate by removing the top section behind the B-post, whereupon skilled coachbuilders hand-formed the lengthened roofline and strut-supported tailgate. With the rear seats folded, 58cu.ft. of luggage space was available; with the seats in the raised position, the car could accomodate passengers in the same manner as the standard saloon, whilst still providing over 35cu.ft. of load carrying space. Standard equipment on these models included an electric sunroof, a vinyl roof covering (in the customers choice of colour), rear seat belts, wash/wipe for the rear window, and central-locking carried over to the large rear door. A fold-away double child seat complete with safety harness, and Dunlop chrome-plated wire wheels were offered at £475.00 and £128.00 respectively, as optional extras.

The Rapport Forte was built in 1981. Based on XJ running gear, the development work had begun around five years earlier. Chris Humberstone was the designer (he was also responsible for the 1973 Owen Sedanca de Ville - there was an obvious likeness) and he envisaged the car being available as a fast-back coupé, a convertible, an estate or a conventional spyder model, with the aid of optional tops.

The front and rear wings were made of Kevlar, with aluminium and light alloys skinning the rest of the structure - for a company better known for doing unthinkable things to Range Rovers, this was quite a hi-tech approach. Had the timing been better, this executive toy might well have taken off. As it was, the company went into the hands of the Receiver, and although about 30 deposits were taken for the Forte, there were only two ever built. One, a brown convertible exists in America, and the other one was on display to the public at the Patrick Collection in Birmingham. The model that belonged to the Patrick Collection was based on V12 mechanicals, and the interior was taken from a Daimler.

The HE Engine

Fears for the future of the V12 were finally put to rest in July 1981, after Jaguar revealed that they had spent half a million pounds in further developing it. The high efficiency (HE) engine's main feature was that of its cylinder head design. The exhaust valve was set

Left:
Ninety years of progress, the 1987 and 1897 Daimlers compared.

recessed below the level of the inlet valve, thus creating a greater turbulence of the fuel/air mixture, which in turn resulted in faster and more efficient combustion. This allowed the engineers to raise the compression ratio from 10:1 to 12.5:1 without any problems, resulting in greater efficiency. A new twin-coil high-energy ignition system was also introduced, and due to the additional torque and power produced by the engine, the final drive ratio was raised from 3.07:1 to 2.88:1, again increasing economy. None of the legendary smoothness was lost and performance was certainly not sacrificed. The resulting benefits were the ability to run on a leaner mixture, cooler running, a less toxic exhaust, and at least a 10% saving on fuel bills. Not surprisingly, this V12 HE engine was still in production over a decade later.

On 15th July, the Double-Six HE was launched at £19,949, with its Vanden Plas counterpart priced at £25,594.40. The electrically-operated seat height adjustment had become a standard fitment on all Daimler models, as had the electrically operated door mirrors and inertia reel rear seat belts. At last, traditional Jaguar themes of value for money and quality workmanship were starting to make a comeback. The *Autocar* published a report on the the Double-Six HE, comparing it to other top executive cars - the Daimler took an easy victory. When *Motor* magazine finally released their report, the testers were equally impressed. They said: *"it is the way that this performance is provided, rather than just cold figures, that makes the Daimler so exceptional... if there is one particular aspect of the Jaguar-Daimler saloons that enables them to hold on to the contentious `Best Car in the World' title, then it must surely be their remarkable and unsurpassed level of refinement."*

The 1983 Face-lift

Many improvements were made to the interior of the cars in 1983. The fascia was given more veneer on the centre console, and this theme was carried over to the door cappings in a similar fashion to the Vanden Plas models. A new thicker rimmed steering wheel was introduced, which featured a larger but neater centre horn push. Outside, the chrome coachline became standard, and the famous front grille was slightly modified to coincide with the discontinuation of the centre bonnet chrome strip. The spoked light alloy wheels, which for so long had been an option, were now a standard fitment, as was a new trip computer.

In the autumn of 1983, the Advanced Jaguar Six engine (the AJ6) was annouced as an optional power-plant in the XJ-S sports car; it was intended that the AJ6 should eventually replace the XK unit. The Sovereign name was taken from Daimler to be used for a limited run of 1,000 top specification Jaguar saloons, after the Sales & Marketing Director of the time, Neil Johnson, had his XJ6 trimmed Daimler style. The new model quickly took off, and now makes up a large percentage of Jaguar sales. Although the Daimlers continued to be the top-of-the-range models, the Sovereign name was never returned to Daimler, and the six cylinder model thereafter was known simply as the 4.2.

The Vanden Plas name was re-allocated, too. It was applied to the North American high specification Jaguars - in effect, that market's equivalent to the Daimler. It seems strange, that one of the oldest names in the motor industry, is still not widely known in America. At a later date, yearly delivery figures were to recognize the Jaguar Vanden Plas models as being Daimlers.

By the end of September 1983, only two saloon models bore the Daimler grille, the 4.2 priced at £21,951 and the Double-Six at £24,991, both offered with automatic transmission only. Daimler's identity of flagship of the range was, however, untarnished, for now fitted as standard to both models was air-conditioning, alloy wheels, cruise control, electric height adjustment on the driver's seat, electrically adjusted door mirrors, electronic stereo radio/cassette, electric sunroof, front fog lights, headlamp wash/wipe, leather trim, rear headrests, rear seat belts and tinted glass all round. A metallic paint finish was a no-cost option, as was a trip computer - only the limited slip differential was listed as a cost option at £174.42, and even that was standard on the Double-Six model.

The Company passed once more into private ownership when it was put on the Stock Market in August 1984, just after massive profits had been reported, and Tom Walkinshaw's team of racing XJ-Ss had won a major event. As the sales figures for the year were released, it was noted that 1984 had been a record year for production. At last, the Jaguar name was again something to be proud of - John Egan had turned the tide. Sadly, 1985 began with the news that Sir William Lyons had died. He was 83, and will go down in

Below: *The Geneva Show 1989. Nearest the camera, a Daimler Double-Six fitted with Ogle style alloy wheels. These are still much more common on Jaguar models, even though they have been a Daimler option since the latter part of 1983.*

Left: *The Vanden Plas Majestic was introduced as the top of the range model for the North American market on June 16th 1989. It features a traditional Daimler style grille, and is fitted with numerous extras throughout the car. Because the Daimler name is, even now, relatively unknown in the USA, the Jaguar Vanden Plas takes the place of the Daimler model.*

motoring history as being one of the industry's most important figures.

Above: *Sir John Egan.*

Project XJ40

October 1986 saw the loss of the 4.2litre model, but a new and exciting vehicle was being developed to take the six cylinder Daimlers into the 1990s - the new XJ6 (known within Jaguar through its development phase as the XJ40). In the meantime, the Series 3 Double-Six was still being listed, now priced at £28,995.

XJ40 prototype schedules had been issued in September 1980, although the project itself had been underway long before this. By the end of the same year, not only had the body structure designs been signed off, but the interior styling had also been completed and approved; the AJ6 engine and GM200R gearbox were also ready for testing. The first of the semi-engineered prototypes was subjected to a 40mph crash test and shortly afterwards, with the completion of sufficient drawings, the first running prototype was finished. As soon as the panel designs were signed off, a glassfibre body was tested in the wind tunnel, fitted with representative underbody parts - all this took about three months. Fully-engineered suspensions (front and back) were completed early in 1981, and the first XJ40 was driven by Jim Randle around the works in July 1981. This gives an indication of how far advanced plans for the project were.

Development continued and, by 1983, prototypes were being tested for durability in places as far afield as the Arizona desert, Nardo in Italy, and in the frozen wastes of Canada. Before the end of the year, a pilot-build facility was constructed, thus allowing more cars to be made for testing. Called SDVs (Specially Designated Vehicles), these cars were to undergo the severe punishment necessary to test the cars to their limit - only then would Jaguar know they had got it right. The preparation that went into XJ40 was extensive; it involved around 400 cars, between them covering millions of miles all over the world, in all conditions.

The AJ6 engine was put into production for use in the XJ-S, and this was to serve as a stringent test of the engines capabilities, putting theory into practice. The AJ6 began life in 1974, and saw many modifications to become the engine that we know today. In 3.6litre form, the AJ6 used four valves per cylinder which were operated by twin overhead camshafts. The Lucas fuel injection system was electronically controlled, and helped the engine to develop a very healthy 221bhp, with maximum torque put at 248lb.ft.

Originally, the gearbox specified as the alternative to the manual Getrag 265 five-speed unit, was the GM200, but because this was taking too long to develop, the Company decided to use the ZF Type 4HP22 automatic transmission instead. To match this new gearbox, a new selector layout, known as the J-gate, was introduced, allowing manual changes to be carried out safely if desired.

The front suspension was basically similar to the previous XJ ranges, but even so, no parts were actually interchangeable. The rear suspension was also well-established in principle, with only detail changes to make this an even better unit in operation, while further reducing the transmission of road and axle noise. Literally hundreds of test rigs (in addition to the worldwide durability tests) made sure that each component used on the vehicle was of the highest quality.

The braking system introduced ABS to the marque for the first time, and the rear disc brakes were now mounted outboard for ease of maintenance. In addition, the wheels and tyres were given a specially designed safety feature, known as Denloc, enabling the bead on the tyre to lock into corresponding grooves on the wheel, giving the driver a better chance of controlling the car in the event of a puncture.

The aerodynamic bodywork had to meet several design needs. Firstly, it had to look new, with no previous Series parts being worked in, but at the same time, it had to retain the Jaguar style. It had to be similar in size to earlier XJ6s, but more spacious inside and also lighter in weight. The car also had to be as quiet and refined as previous models. COB (clear over base) paint methods were to be employed, and from the outset, finish standards and protection from corrosion were to be of utmost concern - all of these points were met admirably.

The interior was completely new, combining traditional leather and burr walnut with the latest in sophisticated electronic driver information (a little too sophisticated as it turned out!). Electronics were also employed, not only for engine management, but also to control the advanced air-conditioning system.

John Egan was given a Knighthood in the summer of 1986. Shortly afterwards, volume production began and the new model took over one of the Browns Lane lines. In August 1986, a Press preview was held at the Institution of Mechanical Engineers in London, with the Press launch being held during the following month on the picturesque roads of Scotland.

The time had arrived at last when, on 8th October 1986, the XJ6 was officially annouced and launched on the UK market. Known simply as the 3.6, the Daimler was still the top model, identifiable by the traditional fluting on the grille and boot plinth, and neutral density rear lamp clusters. The specification was tremendous, and included power assisted steering (with adjustable column), cruise control, ride levelling, limited slip differential, metallic paintwork (with single coachline and chrome body moulding), leather trim, individual rear seats, rear picnic tables, footwell rugs, laminated windscreen, tinted glass, central door locking (which closed windows and sunroof automatically), electric operation of the windows and door mirrors (which were heated), trip computer, heated rear window, courtesy light delay, electrically heated windscreen

Right:
The Daimler 4.0 on display at the 1992 Motor Show, staged at the NEC, Birmingham.

Above:
Nick Scheele, Jaguar's Chairman, with the last Series 3 XJ - a Daimler Double-Six.

Right:
After years of development, 1993 saw the introduction of the long awaited V12 Series 4 (XJ40) - given the Double-Six name, it was a worthy successor to the line, and was being given a different character to the more sporty Jaguar version.

washers, puddle lights, front and rear coat belts, air-conditioning, headlamp power wash, reading lamps, eight-way electrically adjustable heated front seats, rear head rests, electric sunroof, and an electronic audio system complete with six speakers and an electric aerial - all this for £28,495. A manual gearbox was offered as an option, as was a magnolia leather trim or herringbone tweed seat facings. Heated door lock barrels and rear window sunblinds were also offered.

Jaguar had purposely made the engine bay too small for a V engine, as Leyland had wanted to fit the car with a Rover V8. Unfortunately, it caused future engineering problems for Jaguar's own V12, which is naturally larger again. Even though the V12 was not yet available in the new body, this did nothing to dampen the reception that the vehicle received. In all its forms, the XJ has had more accolades bestowed upon it, than any other post-war car in the world, and just one month after the launch, Jim Randle (XJ40s Chief Engineer), received the Top Car 1987 award - proof enough that the new car was a winner. In the next few months, launches for the new model took place all over the world.

In April 1987, the last Series 3 4.2 rolled off the line - only the DS420 limousine remained with the XK engine. The last few 4.2 models (581 of them), had all gone abroad. The delivery figures for 1987 were to show the potential success of the new XJ6; figures of 7,893 sold world wide were certainly impressive. Of that total, about 1,125 XJ6s were sold at home. The Double-Six continued to sell at a reasonable pace, 1,097 were taken worldwide, with 135 of those being retained in the UK.

During the summer of 1989, the new 4.0litre AJ6 engine was introduced. This engine developed 6% more power than the original 3.6litre unit, but more significantly, 13% more torque. To complement this new and much-improved engine, a new four-speed electronically-controlled automatic gearbox was employed, which featureed a switch enabling the driver to choose between Normal and Sport settings. For the time being, the Series 3 Double-Six continued to be the top model, and was priced at £37,000. The Daimler 4.0litre (as it was now known), sold at just £500 less.

Old Names Revived

Later on in 1989, America was given a brand new top-of-the-range machine - the Vanden Plas Majestic. As mentioned earlier, the Vanden Plas name was used in the U.S.A. instead of Daimler, and these VDP cars always carried a Jaguar style grille. The Majestic, however, proudly boasted a fluted radiator shell, and has even more extras fitted to it than the equivalent Daimlers at home. Coachwork was available in any colour, as long as it was red!

At about this time, news of a potential takeover bid was brought to the public's attention. The exchange rate meant that Jaguar's crucial sales in North America had dropped and weakened the Company. Several large concerns were interested in Jaguar, notably General Motors, Ford, BMW and Mercedes-Benz.

At the end of the day, however, Ford managed to gain a controlling interest and took over officially during December 1989. Ironically, Ford had approached Jaguar some years before with a view to producing a badge-engineered car for West Germany and the U.S.A. With Daimler-Benz (Mercedes) in Germany, there would have been little point in marketing a British Daimler there (although a number are sold), so the Lanchester name was very nearly revived. Negotiations broke off at an early stage, although now there is nothing to stop the project going ahead, of course. The only problem is the Lanchester name has been dormant for so long, that market research has shown that very few people associate Lanchester with a car - the average man in the street, unless he was over 50 years old, will probably never even have heard of the genius behind the marque.

However, with Ford now in control, June 1990 saw Sir John Egan leave Jaguar to become the Chief Executive of the BAA. Born in 1939, and educated at Bablake School in Coventry, Egan had saved Jaguar, and in doing so, saved Daimler as well - his great work at the Coventry factory cannot be underestimated. Ford appointed Bill Hayden to take Sir John's place (having arrived at Jaguar in February), and eventually, the job fell to Nick Scheele, following Hayden's renowned *"short, sharp shock treatment."*

Below: *As predicted, the latest Daimler limousine is little more than a stretched XJ40. Built by Eagle Specialist Vehicles, a hearse has also been produced, and the model has the full backing of the factory.*

Left:
The Double-Six name was again retained for the latest V12 Daimler. This is the interior of the car that takes Daimler into its centenary year.

A long wheelbase version of the 4litre XJ40 was announced at the 1992 Motor Show, it having an extra 5 inches in the rear and a slightly raised roofline. The choice of name was quite interesting, the car being christened the "Majestic" for both Jaguar and Daimler versions. The car is finished and trimmed in the old Limousine Shop, but naturally it is not a direct replacement for the "Old Lady".

The development of the V12 powered XJ40 was at last completed, and the car went into production in 1993. Launched at the Geneva Show in March 1993 (the Jaguar XJ12 had been launched in February at Amsterdam), the Daimler version was given the "Double-Six" name, traditionally given to the V12 Daimlers since their first appearance in the 1920s. Powered by a 6litre engine which developed 318bhp at 5,400rpm, the new Double-Six was a far cry from the earlier machines. The GM four speed automatic gearbox was also a great advance, keeping the refinement of the Series 3, whilst offering better performance and economy.

Naturally, this meant that the 5.3litre Series 3 Double-Six, which had been kept in production as a stop-gap until the new car was ready, was finally killed off in December 1992. The last batch of cars were all Daimlers, selling at £43,212. The new car (which sold for £51,700) had a softer character than its Jaguar cousin, the emphasis being on comfort and luxury, but despite this, it was still capable of 0-60mph in less than seven seconds, and had a top speed of 155mph.

In September 1994, the XJ series received its first major face-lift since the introduction of the XJ40 in 1986. The X300 model features a bonnet which is more like a Series 2 or 3 that will doubtless please the enthusiast, and a restyled rear end. Each model has a unique character, and the Daimlers have been given a more supple ride than the Jaguar XJ12, which in turn is less firm than the supercharged XJR (unique to the Jaguar range).

The 4.0litre Daimler is known as the "Six" (at £49,950), while the £59,950 V12 retains the famous Double-Six badge - only a discreet emblem on the boot lid and door pillar tell the two models apart externally. Having driven a new Double-Six as an everyday car for a week, including a 2,000 mile run to Scotland, the author can confirm that the X300 is without doubt one of the finest cars in the world, and will go down in Jaguar history as a landmark for the Company. With effortless performance, exceptional refinement and more than reasonable fuel consumption for a six-litre engine, it is probably the best car ever to wear the exclusive Daimler grille, and is much better than the XJ40 and anything Crewe or Stuttgart currently has to offer.

A new smaller Jaguar-Daimler saloon, the X200, is on the way (by the end of 1998) and there is also talk of a factory designed XJ6 Estate, which should be most interesting - indeed, a prototype seen by the author is a very elegant vehicle. Probably even more exciting is the prospect of a variable torque 4WD saloon - testing has taken place, so let's hope it's just a matter of time before we see it in the showrooms. New engines are also being developed, including a two-stroke V6, and a more conventional V8, the first from the Company since the loss of the Majestic Major unit in 1968.

It is probably just being romantic to think that there could be another unique Daimler in Jaguar's plans, and the chances of a new Lanchester are also very slim - as Andrew Whyte pointed out: *"The marque was misused for far too long by Daimler, which is the main reason why the Lanchester name is not often given its deserved place among the finest cars of its era."* However, the Company has proved that it is proud of its heritage, and indeed, of the Daimler name - why else would they use it for their top of the range models.

The X300 project will easily take the Company into 1996, and so ends the first one-hundred years of the Daimler marque. W.J. Bentley, writing in his book *'Motoring Cavalcade'* noted: *"It is no mere freak of fate that has put the names of Rolls-Royce and Daimler so positively on a pinnacle that nobody could seriously question their pre-eminence."* With any luck, this book will have shown why this statement was made, and why it still holds true.

Right:
X300 was launched in 1994. Basically a new car, there were only fairly minor changes to the styling of the saloons, both at the front end....

Left:
..... and at the rear. Six cylinder and V12 models were available from the outset. For 1996 there will be a special "Century" model.

Appendix I - Pre-War Daimlers.

Year	Type	Hp	Cyl.	B/Stroke	cc	Valve	Ign.	Drive Type/Gears	
1896	4hp	10.0	2	90x120	1527	Poppet	Tube	Chain	4M
1897	4hp	10.0	2	90x120	1527	Poppet	Tube	Chain	4M
1898	4hp	10.0	2	90x120	1527	Poppet	Tube	Chain	4M
	6hp	10.0	2	90x121	1551	Poppet	Tube	Chain	4M
	8hp	20.1	4	90x120	3054	Poppet	Tube	Chain	4M
1899	6hp	10.0	2	90x121	1551	Poppet	Tube	Chain	4M
	Critchley	3.0	2	76x121	1100	Poppet	Tube	Belts	4M
	12hp	20.1	4	90x120	3054	Poppet	Tube	Chain	4M
1900	6/7hp		2			Poppet	Tb/Elec	Prop.	3M
	Kimberley	3.0	2	76x121	1100	Poppet	Tube	Belts	4M
	12hp		4			Poppet	Tb/Elec.	Chain	4M
	16/18hp		4			Poppet	Tb/Elec.	Chain	4M
1901	6/7hp		2			Poppet	Tb/Elec.	Prop.	3M
	Kimberley	3.0	2	76x121	1100	Poppet	Tube	Belts	4M
	8hp	11.0	2	94x130	1804	Poppet	Tb/Elec.	Chain	3M
	12hp	22.4	4	95x120	3402	Poppet	Tb/Elec.	Chain	4M
	16/18hp		4			Poppet	Tb/Elec.	Chain	4M
	24hp		4			Poppet	Tb/Elec.	Chain	4M
1902	8hp	11.0	2	94x130	1804	Poppet	Tb/Elec.	Chain	3M
	12hp	22.4	4	95x120	3402	Poppet	Tb/Elec.	Chain	4M
	22hp	27.3	4	105x130	4503	Poppet	Tb/Elec.	Chain	4M
1903	6/7hp	11.0	2	97x120	1773	Poppet	Bat/Coil	Chain	3M
	12hp	22.4	4	95x120	3402	Poppet	Tb/Elec.	Chain	4M
	14hp	18.3	4	86x100	2324	Poppet	Tb/Elec.	Chain	4M
	22hp	27.3	4	105x130	4503	Poppet	Tb/Elec.	Chain	4M
1904	6/7hp	11.7	2	97x120	1773	Poppet	Bat/Coil	Chain	3M
	16/20hp	20.0	4	90x130	3309	Poppet	Bat/Coil	Chain	4M
	18/22hp	22.4	4	95x135	3827	Poppet	Bat/Coil	Chain	4M
	28/36hp	30.0	4	110x150	5703	Poppet	Bat/Coil	Chain	4M
1905	28/36	30.0	4	110x150	5703	Poppet	Bat/Coil	Chain	4M
	30/40hp	38.1	4	124x150	7247	Poppet	Bat/Coil	Chain	4M
1906	28/36	30.0	4	110x150	5703	Poppet	Bat/Coil	Chain	4M
	30hp	38.1	4	124x150	7247	Poppet	Bat/Coil	Chain	4M
	35hp	44.5	4	134x150	8462	Poppet	Bat/Coil	Chain	4M
	45hp	55.8	4	150x150	10604	Poppet	Bat/Coil	Chain	4M
1907	28hp	35.7	4	120x150	6787	Poppet	Bat/Coil	Chain	4M
	30hp	41.9	4	130x150	7965	Poppet	Bat/Coil	Chain	4M
	35hp	48.6	4	140x150	9237	Poppet	Bat/Coil	Chain	4M
	45hp	55.8	4	150x150	10604	Poppet	Bat/Coil	Chain	4M
1908	30hp	30.0	4	110x130	4942	Poppet	Bat/Coil	Prop.	4M
	36hp	35.7	4	120x150	6787	Poppet	Bat/Coil	Chain	4M
	38hp	38.1	4	124x130	6281	Poppet	Bat/Coil	Prop.	4M
	42hp	41.9	4	130x150	7965	Poppet	Bat/Coil	Chain	4M
	48hp	48.6	4	140x150	9237	Poppet	Bat/Coil	Chain	4M
	58hp	58.8	4	154x140	10431	Poppet	Bat/Coil	Chain	4M
1909	22hp	22.8	4	96x130	3764	Sleeve	Mag.	Prop.	4M
	38hp	38.1	4	124x130	6281	Sleeve	Dual	Prop.	4M
	48hp	48.6	4	140x150	9237	Sleeve	Dual	Chain	4M
	57hp	57.2	6	124x130	9421	Poppet	Dual	Prop.	3M
1910	15hp	15.9	4	80x130	2614	Sleeve	Dual	Prop.	3M
	22hp	22.8	4	96x130	3764	Sleeve	Dual	Prop.	3M
	38hp	38.1	4	124x130	6281	Sleeve	Dual	Prop.	3M
	33hp	34.2	6	96x130	5646	Sleeve	Dual	Prop.	3M
	57hp	57.2	6	124x130	9421	Sleeve	Dual	Prop.	3M
1911	12hp	11.9	4	69x114	1705	Sleeve	Mag.	Prop.	3M
	15hp	15.9	4	80x130	2614	Sleeve	Dual	Prop.	3M
	25hp	25.4	4	101x130	4208	Sleeve	Dual	Prop.	3M
	38hp	38.1	4	124x130	6281	Sleeve	Dual	Prop.	3M
	23hp	23.6	6	80x130	3921	Sleeve	Dual	Prop.	3M
	38hp	38.2	6	101x130	6312	Sleeve	Dual	Prop.	3M
	57hp	57.2	6	124x130	9421	Sleeve	Dual	Prop.	3M
1912	15hp	15.9	4	80x130	2614	Sleeve	Dual	Prop.	3M
	20hp	20.1	4	90x130	3309	Sleeve	Dual	Prop.	4M
	25hp	25.4	4	101x130	4208	Sleeve	Dual	Prop.	4M
	38hp	38.2	4	124x130	6281	Sleeve	Dual	Prop.	4M

Appendix I (continued)

Year	Type	Hp	Cyl.	B/Stroke	cc	Valve	Ign.	Drive Type/Gears	
1912	23hp	23.6	6	80x130	3921	Sleeve	Dual	Prop.	3M
	30hp	30.2	6	90x130	4963	Sleeve	Dual	Prop.	4M
	38hp	38.2	6	101x130	6312	Sleeve	Dual	Prop.	4M
1913	20hp	20.1	4	90x130	3307	Sleeve	Dual	Prop.	4M
	26hp	25.4	4	102x140	4576	Sleeve	Dual	Prop.	4M
	38hp	38.2	4	124x130	6272	Sleeve	Dual	Prop.	4M
	30hp	30.2	6	90x130	4960	Sleeve	Dual	Prop.	4M
	38hp	39.1	6	102x140	6864	Sleeve	Dual	Prop.	4M
1914	20hp	20.1	4	90x130	3307	Sleeve	Mag.	Prop.	3M
	20hp	20.1	4	90x130	3307	Sleeve	Mag.	Prop.	4M
	30hp	30.0	4	110x130	4942	Sleeve	Dual	Prop.	4M
	30hp	30.1	6	90x130	4961	Sleeve	Dual	Prop.	4M
	45hp	45.0	6	110x130	7413	Sleeve	Dual	Prop.	4M
1915	20hp	20.1	4	90x130	3307	Sleeve	Mag.	Prop.	4M
	30hp	30.2	4	110x130	4942	Sleeve	Dual	Prop.	4M
	30hp	30.2	6	90x130	4961	Sleeve	Dual	Prop.	4M
	45hp	45.0	6	110x130	7413	Sleeve	Dual	Prop.	4M
1919	30hp	30.2	6	90x130	4962	Sleeve	Mag.	Prop.	4M
	30hp	30.2	6	90x120	4962	Sleeve	Mag.	Prop.	4M
	45hp	45.0	6	110x130	7413	Sleeve	Mag.	Prop.	4M
1920	30hp	30.5	6	90x130	4962	Sleeve	Mag.	Prop.	4M
	30hp	30.2	6	90x120	4962	Sleeve	Mag	Prop.	4M
	45hp	45.0	6	110x130	7413	Sleeve	Mag.	Prop.	4M
1921	30hp	30.1	6	90x130	4962	Sleeve	Mag.	Prop.	4M
	30hp	30.1	6	90x130	4962	Sleeve	Mag.	Prop.	4M
	45hp	45.0	6	110x130	7413	Sleeve	Mag.	Prop.	4M
1922	20hp	20.1	4	90x130	3308	Sleeve	Mag.	Prop.	4M
	30hp	30.1	6	90x130	4962	Sleeve	Mag.	Prop.	4M
	30hp	30.1	6	90x130	4962	Sleeve	Mag.	Prop.	4M
	45hp	45.0	6	110x130	7413	Sleeve	Mag.	Prop.	4M
1923	12hp	12.9	6	59x94	1542	Sleeve	Coil	Prop.	4M
	16hp	16.5	6	66x104	2167	Sleeve	Coil	Prop.	4M
	21hp	20.9	6	75x114	3021	Sleeve	Coil	Prop.	4M
	30hp	30.1	6	90x130	4962	Sleeve	Mag.	Prop.	4M
	45hp	45.0	6	110x130	7413	Sleeve	Mag.	Prop.	4M
1924	16hp	15.7	6	65x94	1872	Sleeve	Coil	Prop.	4M
	20hp	20.1	6	74x104	2648	Sleeve	Coil	Prop.	4M
	20hp	20.1	6	74x104	2648	Sleeve	Coil	Prop.	4M
	21hp	20.9	6	75x114	3022	Sleeve	Coil	Prop.	4M
	25hp	24.7	6	81x130	3568	Sleeve	Coil	Prop.	4M
	30hp	30.1	6	90x130	4962	Sleeve	Mag.	Prop.	4M
	35hp	35.0	6	97x130	5764	Sleeve	Coil	Prop.	4M
	45hp	45.0	6	110x130	7413	Sleeve	Mag.	Prop.	4M
1925	16hp	15.7	6	65x94	1872	Sleeve	Dual	Prop.	3M
	20hp	20.1	6	74x104	2648	Sleeve	Dual	Prop.	4M
	21hp	20.9	6	75x114	3022	Sleeve	Dual	Prop.	4M
	25hp	24.7	6	81x114	3568	Sleeve	Dual	Prop.	4M
	30hp	30.1	6	90x130	4962	Sleeve	Dual	Prop.	4M
	35hp	35.0	6	97x130	5764	Sleeve	Dual	Prop.	4M
	45hp	45.0	6	110x130	7413	Sleeve	Dual	Prop.	4M
	45hp	51.4	6	117x130	8458	Sleeve	Dual	Prop.	4M
1926	16/55	15.7	6	65x94	1872	Sleeve	Dual	Prop.	3M
	20/70	20.1	6	74x104	2648	Sleeve	Dual	Prop.	4M
	25/85	24.7	6	81x114	3568	Sleeve	Dual	Prop.	4M
	35/120	35.0	6	97x130	5764	Sleeve	Dual	Prop.	4M
	30hp	35.0	6	97x130	5764	Sleeve	Dual	Prop.	4M
	45hp	51.4	6	117x130	8458	Sleeve	Dual	Prop.	4M
1927	16/55	15.7	6	65x94	1872	Sleeve	Dual	Prop.	3M
	20/70	20.1	6	74x104	2648	Sleeve	Dual	Prop.	4M
	25/85	24.7	6	81x114	3568	Sleeve	Dual	Prop.	4M
	35/120	35.0	6	97x130	5764	Sleeve	Dual	Prop.	4M
	50hp	49.4	12	81x114	7136	Sleeve	Dual	Prop.	4M
1928	16/55	15.7	6	65x94	1872	Sleeve	Dual	Prop.	3M
	20/70	20.1	6	74x104	2648	Sleeve	Dual	Prop.	4M
	25/85	24.7	6	81x114	3568	Sleeve	Dual	Prop.	4M
	35/120	35.0	6	97x130	5764	Sleeve	Dual	Prop.	4M
	30hp	31.4	12	65x94	3744	Sleeve	Dual	Prop.	4M
	50hp	49.4	12	81x114	7136	Sleeve	Dual	Prop.	4M

Appendix I (continued)

Year	Type	Hp	Cyl.	B/Stroke	cc	Valve	Ign.	Drive Type	Gears
1929	16/55	15.7	6	65x94	1872	Sleeve	Coil	Prop.	3M
	20/70	20.1	6	74x104	2648	Sleeve	Coil	Prop.	4M
	25/85	24.7	6	81x114	3568	Sleeve	Dual	Prop.	4M
	35/120	35.0	6	97x130	5764	Sleeve	Dual	Prop.	4M
	30hp	31.4	12	65x94	3744	Sleeve	Dual	Prop.	4M
	50hp	49.4	12	81x114	7136	Sleeve	Dual	Prop.	4M
1930	20hp	20.1	6	74x104	2648	Sleeve	Coil	Prop.	4M
	25hp	24.7	6	81x114	3568	Sleeve	Coil	Prop.	4M
	35hp	35.0	6	97x130	5764	Sleeve	Coil	Prop.	4M
	30hp	31.4	12	65x94	3744	Sleeve	Coil	Prop.	4M
	50hp	49.4	12	81x114	7136	Sleeve	Coil	Prop.	4M
1931	20hp	20.1	6	74x104	2648	Sleeve	Coil	Prop.	4M
	20/30	24.7	6	81x114	3568	Sleeve	Coil	Prop.	4M/P
	35hp	35.0	6	97x130	5764	Sleeve	Coil	Prop.	4M/P
	30hp	31.4	12	65x94	3744	Sleeve	Coil	Prop.	4M/P
	30/40	40.2	12	74x104	5296	Sleeve	Coil	Prop.	4M/P
	40/50	49.4	12	81x104	6511	Sleeve	Coil	Prop.	4M/P
1932	16/20	20.1	6	74x104	2648	Sleeve	Coil	Prop.	4P
	20/25	24.8	6	81x114	3568	Sleeve	Coil	Prop.	4P
	35hp	35.0	6	97x130	5764	Sleeve	Coil	Prop.	4P
	30hp	31.4	12	65x94	3744	Sleeve	Coil	Prop.	4P
	30/40	40.2	12	74x104	5296	Sleeve	Coil	Prop.	4P
	40/50	49.4	12	81x104	6511	Sleeve	Coil	Prop.	4P
1933	15hp	15.0	6	64x90	1805	Poppet	Coil	Prop.	4P
	20hp	20.1	6	74x104	2648	Sleeve	Coil	Prop.	4P
	25hp	24.7	6	81x114	3568	Sleeve	Coil	Prop.	4P
	40hp	40.2	12	74x104	5296	Sleeve	Coil	Prop.	4P
	50hp	49.4	12	81x104	6511	Sleeve	Coil	Prop.	4P
1934	15hp	15.0	6	64x90	1805	Poppet	Coil	Prop.	4P
	20hp	19.3	6	72x110	2700	Poppet	Coil	Prop.	4P
	25hp	24.7	6	81x114	3568	Sleeve	Coil	Prop.	4P
	40hp	40.2	12	74x104	5296	Sleeve	Coil	Prop.	4P
	50hp	49.4	12	81x104	6511	Sleeve	Coil	Prop.	4P
1935	15hp	15.0	6	64x105	2003	Poppet	Coil	Prop.	4P
	20hp	19.3	6	72x110	2687	Poppet	Coil	Prop.	4P
	25hp	25.7	8	72x115	3746	Poppet	Coil	Prop.	4P
	40hp	40.2	12	74x104	5296	Sleeve	Coil	Prop.	4P
	50hp	49.4	12	81x104	6511	Sleeve	Coil	Prop.	4P
1936	15hp	15.0	6	64x105	2003	Poppet	Coil	Prop.	4P
	Lgt 20	19.3	6	72x105	2565	Poppet	Coil	Prop.	4P
	20hp	23.8	6	80x110	3317	Poppet	Coil	Prop.	4P
	Lgt Str 8	25.7	8	72x105	3421	Poppet	Coil	Prop.	4P
	Str 8	31.7	8	80x115	4624	Poppet	Coil	Prop.	4P
1937	15hp	16.2	6	66x105	2166	Poppet	Coil	Prop.	4P
	20hp	19.3	6	72x105	2565	Poppet	Coil	Prop.	4P
	24hp	23.8	6	80x110	3317	Poppet	Coil	Prop.	4P
	Lgt Str 8	25.7	8	72x105	3421	Poppet	Coil	Prop.	4P
	Str 8	31.7	8	80x115	4624	Poppet	Coil	Prop.	4P
	DD6	49.4	12	81x104	6511	Poppet	Coil	Prop.	4P
1938	15hp	16.2	6	66x105	2166	Poppet	Coil	Prop.	4P
	20hp	19.3	6	72x105	2565	Poppet	Coil	Prop.	4P
	24hp	23.8	6	80x110	3317	Poppet	Coil	Prop.	4P
	Lgt Str 8	25.7	8	72x105	3421	Poppet	Coil	Prop.	4P
	Str 8	31.7	8	80x115	4624	Poppet	Coil	Prop.	4P
1939	15hp	18.0	6	70x110	2522	Poppet	Coil	Prop.	4P
	20hp	19.3	6	72x105	2565	Poppet	Coil	Prop.	4P
	24hp	23.8	6	80x110	3317	Poppet	Coil	Prop.	4P
	Str 8	29.8	8	77x105	3960	Poppet	Coil	Prop.	4P
	Str 8	31.7	8	80x115	4624	Poppet	Coil	Prop.	4P
1940	15hp	18.0	6	70x110	2522	Poppet	Coil	Prop.	4P
	20hp	19.3	6	72x105	2565	Poppet	Coil	Prop.	4P
	24hp	23.8	6	80x110	3317	Poppet	Coil	Prop.	4P
	Str 8	29.8	8	77x105	3960	Poppet	Coil	Prop.	4P
	Str 8	31.7	8	80x115	4624	Poppet	Coil	Prop.	4P

Key: Hp - Treasury Rating
4M - Four-speed manual gearbox.
4M/P- Four-speed manual or pre-select gearbox.
4P - Four-speed pre-select/fluid-flywheel only.

Appendix II - Pre-War Daimler Chassis Types.

1902	TA.12	TA.22					
1903	TA.6	TA.12	TB.14	TB.22	TC.22		
1904	TA.18	TA.28	TB.18	TB.28	TC.16	TC.28	
1905	TA.17	TA.33	TB.17	TC.17	TD.30	TD.35	TD.28
	TE.28	TG.28	CA.28				
1906	TA.45	TH.28	TH.30	TH.35	TJ.28	TJ.30	TJ.35
	TK.28	TK.30	CA.16	CB.28			
1907	TD.17	TL.28	TL.30	TL.35	TL.45		
	TM.28	TM.30	TM.35	TM.45	TO.28	TO.30	TO.35
	MA.28	KP.45					
1908	TA.30	TA.38	TB.42	TB.48	TB.58		
	TC.42	TC.48	TC.58	TD.58	FA.15		
1909	TB.22	TB.38	TC.38	TC.48	TC.57	TC.22	
1910	TB.15	TC.33	TD.22	TD.33	TD.38	TD.57	
	TE.22	TE.38	TE.57	TE.33	TF.33	FA.15	15SS
1911	TA.23	TA.25	TB.23	TB.25	TC.15	TD.15	TG.38
	TH.38	TJ.38	TK.38	TA6.38	TB6.38		
1912	TAC6.38	TB.20	TB6.30	TBD6.38	TC.20	TC.25	TC6.38
	TD.25	TE.15	TG.15	TL.38	TM.38	TO6.38	
1913	TA.26	TA6.38	TB.26	TB6.38	TE.20	TE.30	TG.20
	TH.15	TH.20	TJ.15	TO.20	TO.30	TR.38	TS.38
1914	TB4.30	TB6.45	TC4.30	TC6.45	TG6.30	TH6.30	TJ.20
	TK.20	TL.20	TM.20	CC.30	CD.30	CB.22	
1915	TOP.20						
1916	TRS.20						
1919	TGH.45	TLM.30	TOP.30	CJK.22			
1921	TT.20						
1922	TB.21	TC.21	TJ.45	TS.30			
1923	TB.12	TB.16	TC.16				
1924	C.35	C.25	C.20	D.16	D.20		
1925	A.45	A1.21	A2.21	L1.16	L4.16	N.30	N.45
	R.30	R.45	S.20	S.25	S.35	T.20	T.25
	T.35						
1926	L5.16	N.35/120	R.35/120	S.20/70	S.20/85	T.20/70	T25/85

Appendix III - Pre-War Daimler Chassis Numbers.

1933	15hp	33700 - 34354	15hp	34355 - 35202
	15hp	35208 - 35707	LQ2/20	36000 - 36049
	LQ2/20	36150 - 36199		
1934	15hp	37000 - 37999	LQ2/20	36200 - 36299
	LQ2/20	38000 - 38049	LQ2/20	38050 - 38149
	V26	36100 - 36107	V26	36958 - 36999
	V26	38250 - 38299		
1935	15hp	38500 - 39145	15hp	39150 - 39649
	15hp	39850 - 40099	15hp	40200 - 40449
	LQ2/20	38150 - 38249	LQ3/20	36109-36116
	LQ3/20	36117 - 36149	LQ3/20	36300 - 36309
	LQ3/20	36600 - 36649	LQ3/20	36650 - 36869
	V26	38300 - 38449		
1936	15hp	40450 - 40949	15hp	41350 - 42349
	E.20	41050 - 41249	E.20	42350 - 42551
	LQ3/24	40950 - 41049	LQ3/24	42600 - 42649
	E.3½	40100 - 40199	V.4½	41250 - 41349
	V&F.4½	43500 - 43624		
1937	15/17	43000 - 43499	15/17	43850 - 44199
	15/17	44300 - 44499	15/17	45800 - 46099
	DB.17/1	44500 - 45499	DB.17/2	48550 - 49049
	E.20	42650 - 42974	E.20	47600 - 47749
	EL.24	43725 - 43849	E.3½	43625 - 43674
	V&F.4½	44200 - 44299	V&F.4½	48300 - 48549
1938/9	E.4ltr	47781 - 47899	EL.24	45500 - 45674
	EL.24	47900 - 48273	ELS.24	48274 - 48298
	E.3½	47750 - 47780	DB18/1	46100 - 47599

Appendix IV - Post-War Daimlers.

All of the post-war Daimlers are listed here (Jaguar Daimlers see appendix VI)

Year	Chassis Type	Model	No. Cyls.	Bore& Stroke	c c
1946	DB.18		6	70x110	2522
	DE.27		6	85x120	4095
	DE.36	Straight-Eight	8	85x120	5460
1947	DB.18		6	70x110	2522
	DE.27		6	85x120	4095
	DE.36	Straight-Eight	8	85x120	5460
1948	DB.18		6	70x110	2522
	DB.18	Special Sports	6	70x110	2522
	DE.27		6	85x120	4095
	DE.36	Straight-Eight	8	85x120	5460
1949	DB.18		6	70x110	2522
	DB.18	Special Sports	6	70x110	2522
	DE.27		6	85x120	4095
	DE.36	Straight-Eight	8	85x120	5460
1950	DB.18		6	70x110	2522
	DB.18	Special Sports	6	70x110	2522
	DB.18	Consort	6	70x110	2522
	DC.27	Ambulance	6	85x120	4095
	DE.27		6	85x120	4095
	DH.27		6	85x120	4095
	DE.36	Straight-Eight	8	85x120	5460
1951	DB.18	Special Sports	6	70x110	2522
	DB.18	Consort	6	70x110	2522
	DB.18	Empress	6	70x110	2522
	DC.27	Ambulance	6	85x120	4095
	DE.27		6	85x120	4095
	DE.36	Straight-Eight	8	85x120	5460
	DF.300/1	Regency	6	76x108	2952
1952	DB.18	Special Sports	6	70x110	2522
	DB.18	Consort	6	70x110	2522
	DC.27	Ambulance	6	85x120	4095
	DE.36	Straight-Eight	8	85x120	5460
	DF.300/1	Regency	6	76x108	2952
	DF.302/3	Empress II	6	76x108	2952
	DF.302/3	Conv. Coupe	6	76x108	2952
1953	DB.18	Special Sports	6	70x110	2522
	DB.18	Consort	6	70x110	2522
	DC.27	Ambulance	6	85x120	4095
	DE.36	Straight-Eight	8	85x120	5460
	DF.300/1	Regency	6	76x108	2952
	DF.302/3	Empress II	6	76x108	2952
	DF.302/3	Conv. Coupe	6	76x108	2952
	DJ.250/1	Conquest	6	76x89	2433
	DJ.254/5	Roadster	6	76x89	2433
1954	DC.27	Ambulance	6	85x120	4095
	DF.304/5	Regency II	6	83x108	3468
	DF.306/7	Empress IIA	6	83x108	3468
	DF.308/9	Reg. Sportsman	6	83x108	3468
	DF400/1		6	95x108	4617
	DJ.250/1	Conquest	6	76x89	2433
	DJ.252/3	Conquest dhc	6	76x89	2433
	DJ.254/5	Roadster	6	76x89	2433
	DJ.256/7	Conq Century	6	76x89	2433
	DK.400		6	95x108	4617
1955	DF.306/7	Empress IIA	6	83x108	3468
	DF.310/1	One-O-Four	6	83x108	3468
	DF.310/1	Four-Light	6	83x108	3468
	DF.310/1	Ladies Model	6	83x108	3468
	DF.402/3		6	95x108	4617
	DJ.250/1	Conquest II	6	76x89	2433
	DJ.254/5	New dhc	6	76x89	2433
	DJ.256/7	Century II	6	76x89	2433
	DK.400		6	95x108	4617
1956	DF.308/9	Empress III	6	83x108	3468
	DF.310/1	One-O-Four	6	83x108	3468
	DF.310/1	Ladies Model	6	83x108	3468
	DF.314	Auto "104"	6	83x108	3468
	DF.402/3		6	95x108	4617
	DJ.250/1	Conquest II	6	76x89	2433
	DJ.254/5	New dhc	6	76x89	2433
	DJ.256/7	Century II	6	76x89	2433
	DJ.260	Auto Century	6	76x89	2433
	DK.400		6	95x108	4617
1957	DF.308/9	Empress III	6	83x108	3468
	DF.310/1	One-O-Four	6	83x108	3468
	DF.310/1	Ladies Model	6	83x108	3468
	DJ.254/5	New dhc	6	76x89	2433
	DJ.256/7	Century II	6	76x89	2433
	DJ260	Auto Century	6	76x89	2433
	DK.400		6	95x108	4617

Appendix IV (continued)

Year	Chassis Type	Model	No. Cyls.	Bore& Stroke	c c
1958	DF.316/7	Majestic	6	86x108	3794
	DJ.256/7	Century II	6	76x89	2433
	DK.400		6	95x108	4617
1959	DF.316/7	Majestic	6	86x108	3794
	DK.400		6	95x108	4617
	SP.250		8	76x70	2548
1960	DF.318/9	Majestic	6	86x108	3794
	DQ.450/1	M-Major	8	95x80	4561
	SP.250		8	76x70	2548
1961	DF.318/9	Majestic	6	86x108	3794
	DQ.450/1	M-Major	8	95x80	4561
	DR.450/1	M-Major	8	95x80	4561
	SP.250		8	76x70	2548
1962	DF.318/9	Majestic	6	86x108	3794
	DQ.450/1	M-Major	8	95x80	4561
	DR.450/1	M-Major	8	95x80	4561
	SP.250		8	76x70	2548
1963	DQ.450/1	M-Major	8	95x80	4561
	DR.450/1	M-Major	8	95x80	4561
	SP.250		8	76x70	2548
1964	DQ.450/1	M-Major	8	95x80	4561
	DR.450/1	M-Major	8	95x80	4561
	SP.250		8	76x70	2548
1965	DQ.450/1	M-Major	8	95x80	4561
	DR.450/1	M-Major	8	95x80	4561
1966	DQ.450/1	M-Major	8	95x80	4561
	DR.450/1	M-Major	8	95x80	4561
1967	DQ.450/1	M-Major	8	95x80	4561
	DR.450/1	M-Major	8	95x80	4561
1968	DQ.450/1	M-Major	8	95x80	4561
	DR.450/1	M-Major	8	95x80	4561

Appendix V - Post-War Daimler Chassis Numbers.

This is a list of consecutive chassis numbers for all post-war Daimlers up to those with Jaguar-type references. In some cases, although chassis numbers were allocated, it does not always follow that this amount of cars were built.

50000 - 50005	DE.27	50006 - 50011	DE.36
50013 - 50031	DB.18	50040 - 51039	DB.18
51040 - 51133	DE.27	51150 - 51243	DE.36
51250 - 51699	DE.27	51700 - 51749	DE.36
51750 - 51759	DE.36(Lhd)	51800 - 52799	DB.18
52800 - 52825	DE.36	52900 - 52949	DH.27
52950 - 52959	DH.27	53000 - 53749	DB.18
53750 - 53999	Special Sports	54000 - 54499	DC.27
54500 - 54949	DB.18	54950 - 54993	DE.36(Lhd)
54994 - 54999	DE.36(Lhd)	55000 - 55999	Consort
56000 - 56135	DB.18	56136 - 56188	Special Sports
56200 - 56699	Consort	56700 - 56774	Special Sports
57000 - 57007	Three-litre	57010 - 59009	Consort
59010 - 59236	Special Sports	59240 - 59988	Consort
80000 - 81999	DF.300/1	82000 - 82027	DF.302
82400 - 82404	DF.303	82500 - 84499	DJ.250(Basic)
84500 - 84540	DJ.251(Basic)	85050 - 87549	DJ.250/1
87550 - 87749	DJ.252	87750 - 87799	DJ.253
88000 - 88010	DF.304/5	88250 - 88633	DF.304/5
88700 - 89057	DF.310/1	90000 - 90016	One-O-Four
90100 - 90234	DF.314	90450 - 90497	DJ.254/5
90500 - 90553	DJ.254/5 Dhc	90850 - 90866	DJ.254/5 Roadster
90950 - 91449	DJ.256/7	91450 - 91475	DF.306/7
91476 - 91507	DF.308/9	91700 - 92699	DJ.256/7
92700 - 92724	DK.400	92725 - 92785	DF.400/1/2
92965 - 92969	DF.403	92975 - 93974	DJ.256/7
95000 - 95999	DJ.256/7	96000 - 96106	DK.400
96500 - 96849	DJ.256/7	96850 - 96898	Ladies Model
97050 - 97517	DJ.256/7	97550 - 98049	DJ.260/1
98050 - 98299	DJ.260/1	98300 - 98749	Majestic(Rhd)
98750 - 98769	Majestic(Lhd)	98770 - 99774	Majestic(Rhd)
99780 - 99794	Majestic(Lhd)	100000 - 100005	SP.250
100010 - 100569	SP.250(Lhd)	100570 - 100759	SP.250(Rhd)
100760 - 101324	SP.250(Lhd)	101325 - 101509	SP.250(Rhd)
101510 - 101585	SP.250(Lhd)	102510 - 102834	SP.250(Rhd)
103710 - 104456	SP.250(Rhd)	136001 - 136002	DR.450
136011 - 136699	DR.450/1	136701 - 136710	DQ.450
136711 - 136890	DQ.450	136891 - 136915	DQ.451
136916 - 137891	DQ.450	139001 - 139176	DR.450/1

Appendix VI - The Daimler by Jaguar.

Year	Model	cc	Chassis No. (Rhd)	(Lhd)
Eight-cylinder cars:				
1962-67	2½ Litre V8	2.5	1A.1001 - 13377 (R)	1A.20001 - 20622 (L)
1967-69	V8-250	2.5	1K.1001 - 5780 (R)	1K.30001 - 30105 (L)
Six-cylinder cars:				
1966-69	420 Sovereign	4.2	1A.30001-35476(R)	1A.70001 - 70355 (L)
1969-73	Ser.1 Sovereign	2.8	1T.1001 - 4069 (R)	1T.50001 - 50165 (L)
1969-73	Ser.1 Sovereign	4.2	1U.1001 - 11894 (R)	1U.50001 - 50726 (L)
1972-73	Ser.1 SovLWB	4.2	2D.1001 - 1394 (R)	2D.50001 (L)
1975-78	Ser.2 Sovereign	3.4	3B.1001 - 3345 (R)	3B.50001 - 50004 (L)
1973-74	Ser.2 Sovereign	4.2	2M.1001 - 3313 (R)	2M.50001 - 50115 (L)
1973-78	Ser.2 SovLWB	4.2	2S.1001 - 12816 (R)	2S.50001 - 50704 (L)
1975-78	Ser2 Sov 2dr	4.2	2H.1001 - 2586 (R)	2H50001 - 50112(L)
1975-78	Ser2 4.2 VDP	4.2	3C.1001 - 1818 (R)	3C.50001 - 50012 (L)
1978-79	Ser.2 SovLWB	4.2	100001 - 111737	
1978-79	Ser.2 4.2 VDP	4.2	100001 - 111737	
1979-83	Ser.3 Sovereign	4.2	300001 - 487641	
1979-83	Ser.3 4.2 VDP	4.2	300001 - 487641	
1983-86	Ser.3 Daim 4.2	4.2	300001 - 487641	
1986-89	XJ40 Daim 3.6	3.6	500001 - 708757	
1989-94	XJ40 Daim 4.0	4.0	500001 - 708757	
1994-	X300 Six	4.0	720000 -	
1968-78	DS.420	4.2	1M.1001 - 3616 (R)	1M.20001 - 20151 (L)
1978-79	DS.420	4.2	10001 - 10209	
1979-93	DS.420	4.2	200001 - 201631	
1979-93	DS D/A Ch.	4.2	400001 - 400425	
Twelve-cylinder cars:				
1972-73	Ser.1 DD6	5.3	2A.1001 - 1524 (R)	2A.50001 - 50011 (L)
1972-73	Ser.1 DD6 VDP	5.3	2B.1001 - 1337 (R)	2B.50001 - 50005 (L)
1973-78	Ser.2 DD6	5.3	2K.1001 - 2925 (R)	2K.50001 - 50227 (L)
1975-78	Ser.2 DD6 2dr	5.3	2F.1001 - 1372 (R)	2F.50001 - 50027 (L)
1973-78	Ser.2 DD6 VDP	5.3	2P.1001 - 2220 (R)	2P.50001 - 50355 (L)
1978-79	Ser.2 DD6	5.3	100001 - 111737	
1978-79	Ser.2 DD6 VDP	5.3	100001 - 111737	
1979-81	Ser.3 DD6	5.3	300001 - 487641	
1979-83	Ser.3 DD6 VDP	5.3	300001 - 487641	
1981-93	Ser.3 DD6 HE	5.3	300001 - 487641	
1993-94	XJ40 DD6	6.0	500001 - 708757	
1994-	X300 DD6	6.0	720000 -	

Note: VIN numbers unfortunately tell the historian nothing without the code letters. For instance, using just numbers, a late Series 2 saloon could just as easily be an early XJ-S, Lhd or Rhd. Nor is there any distinction between Jaguar and Daimler models. Numbers alone can therefore only provide a guide.

Appendix VII - Strict Lanchesters.

This is a list of all Lanchester cars built before 1932 model year.

Year	Type	Hp	No. Cyls.	Bore & Stroke	cc	Ign.	Drive Type/Gears	
1895	5 hp		1	4.75 x 4.5	1,304	Coil	Chains	2M
1897	8 hp		2	5 x 5.75	3,697	Magneto	Worm	2M
1898	8 hp		2	5 x 5.75	3,697	Magneto	Worm	2M
1900	10hp	22	2	5.25 x 5.69	4,031	Magneto	Worm	3M
1901	10hp	22	2	5.25 x 5.69	4,031	Magneto	Worm	3M
1902	12hp	22	2	5.25 x 5.69	4,031	Magneto	Worm	3M
1903	12hp	22	2	5.25 x 5.69	4,031	Magneto	Worm	3M
	16hp		2	5.5 x 5.69	4,428	Magneto	Worm	3M
1904	12hp	22	2	5.25 x 5.69	4,031	Magneto	Worm	3M
	18hp		2	5.75 x 5.69	4,836	Magneto	Worm	3M
	20 hp	26	4	4 x 3	2,470	Magneto	Worm	2M
1905	12hp	22	2	5.25 x 5.69	4,031	Magneto	Worm	3M
	20 hp	26	4	4 x 3	2,470	Magneto	Worm	3M
1906	20 hp	26	4	4 x 3	2,470	Magneto	Worm	3M
	28 hp	39	6	4 x 3	3,705	Magneto	Worm	3M
1907	20 hp	26	4	4 x 3	2,470	Magneto	Worm	3M
	28 hp	39	6	4 x 3	3,705	Magneto	Worm	3M
1908	20 hp	26	4	4 x 3	2,470	Magneto	Worm	3M
	28 hp	39	6	4 x 3	3,705	Magneto	Worm	3M
1909	20 hp	26	4	4 x 3	2,470	Magneto	Worm	3M
	28 hp	39	6	4 x 3	3,705	Magneto	Worm	3M
1910	20 hp	26	4	4 x 3	2,470	Magneto	Worm	3M
	28 hp	39	6	4 x 3	3,705	Magneto	Worm	3M
	38 hp	39	6	4 x 4	4,940	Dual	Worm	3M
1911	25 hp	26	4	4 x 4	3,293	Dual	Worm	3M
	38 hp	39	6	4 x 4	4,940	Dual	Worm	3M
1912	25 hp	26	4	4 x 4	3,293	Dual	Worm	3M
	38 hp	39	6	4 x 4	4,940	Dual	Worm	3M
1913	25 hp	26	4	4 x 4	3,293	Dual	Worm	3M
	38 hp	39	6	4 x 4	4,940	Dual	Worm	3M
1914	25 hp	26	4	4 x 4	3,293	Dual	Worm	3M
	38 hp	39	6	4 x 4	4,940	Dual	Worm	3M
	40 hp	39	6	4 x 4.5	5,542	Dual	Worm	3M
1915	AFV	39	6	4 x 4	4,940	Dual	Worm	3M
1916	AFV	39	6	4 x 4	4,940	Dual	Worm	3M
1919	38hp	39	6	4 x 5	6,174	Dual	Worm	3M
1920	38hp	39	6	4 x 5	6,174	Dual	Worm	3M
1921	38hp	39	6	4 x 5	6,174	Dual	Worm	3M
1922	38hp	39	6	4 x 5	6,174	Dual	Worm	3M
1923	21hp	21	6	76.5 x 114	3,142	Magneto	Worm	4M
	38hp	39	6	4 x 5	6,174	Dual	Worm	3M
1924	21hp	21	6	76.5 x 114	3,142	Magneto	Worm	4M
	38hp	39	6	4 x 5	6,174	Dual	Worm	3M
1925	21hp	21	6	78.7 x 114	3,327	Magneto	Worm	4M
	38hp	39	6	4 x 5	6,174	Dual	Worm	3M
	AFV	39	6	4 x 5	6,174	Dual	Worm	6M
1926	21hp	21	6	78.7 x 114	3,327	Magneto	Worm	4M
	38hp	39	6	4 x 5	6,174	Dual	Worm	3M
	AFV	39	6	4 x 5	6,174	Dual	Worm	6M
1927	21hp	21	6	78.7 x 114	3,327	Magneto	Worm	4M
	38hp	39	6	4 x 5	6,174	Dual	Worm	3M
	AFV	39	6	4 x 5	6,174	Dual	Worm	6M
1928	21hp	21	6	78.7 x 114	3,327	Magneto	Worm	4M
	30hp	31	8	78.7 x 114	4,440	Dual	Worm	4M
	AFV	39	6	4 x 5	6,174	Dual	Worm	6M
1929	21hp	21	6	78.7 x 114	3,327	Magneto	Worm	4M
	30hp	31	8	78.7 x 114	4,440	Dual	Worm	4M
	AFV	39	6	4 x 5	6,174	Dual	Worm	6M
1930	21hp	21	6	78.7 x 114	3,327	Magneto	Worm	4M
	30hp	31	8	78.7 x 114	4,440	Dual	Worm	4M
	AFV	39	6	4 x 5	6,174	Dual	Worm	6M
1931	15/18	18	6	69.5 x 110	2,504	Coil	Worm	4P
	30hp	31	8	78.7 x 114	4,440	Dual	Worm	4M

Note: Bore & Stroke - pre-1920 cars in inches, new post-1920 cars in millimetres. Armoured Fighting Vehicles always in inches.

Left: *An 1897 Daimler pictured alongside its modern counterpart.*

Appendix VIII - The Lanchester by Daimler.

All of the Daimler designed and built Lanchesters are listed below, along with their year of manufacture and chassis numbers.

Year	Model	Chassis No.	Model	Chassis No.
1933	10hp	15000 - 18038	10hp	18059 - 19555
	10hp	19556 - 20554 (inc. 1932 cars)	15/18	10000 - 10800
1934	10hp	20555 - 23055	10hp	23056 - 24056
	18hp	10801 - 11752	18hp	11754 - 12037
1935	10hp	24057 - 25055	10hp	25062 - 25311
	12hp	26000 - 26502	18hp	12291 - 12649
1936	10hp	28000 - 28999	10hp	29500 - 29999
	12hp	26550 - 26974	12hp	29000 - 29148 (De Luxe)
	E.18	12650 - 12841	E.18	13200 - 13222
1937	11hp	32000 - 32999	14hp	30000 - 31999 (Roadrider)
	14hp	33000 - 33424 (Roadrider)	E.18	13275 - 13624
1938	11hp	35425 - 36424	14-2	33425 - 35024 (De Luxe)
	14-3	35025 - 35124 (Export)		
1939	14-2	35125 - 35274	14-3	35025 - 35124
1946/51	LD.10			60000 - 60024 (P/types)
		60025 - 61074 (Briggs)		61800 - 62799 (Briggs)
		62800 - 63225 (Briggs)		63226 - 63804 (Barker)
1951/4	LJ.200/1			65007 - 65999 (Rhd)
		66000 - 66315 (Lhd)		67150 - 67943 (Rhd)
1953	LJ.252			70006 - 70007 (Dauphin)
1954/6	LM.150/1			70100 - 70112 (Sprite)

Appendix IX - BSA Car Production

A listing of all the known BSA cars, Daimler/Lanchester-built or otherwise.

Dates	Type	Engine	Notes
1907-1910	18/23 hp	2.5 litre 4-cyl. (L-head)	BSA's first car. Late to make it into production.
1909-1910	14/18hp	4.2 litre 4-cyl. (L-head)	Larger version of above.
1909-1910	25/33hp	5.4 litre 4-cyl. (T-head)	Closely based on Itala design.
1912-1915	BSA	12 hp Daimler sleeve-valve	Similiar to early Stoneleigh. Body of pressed steel built by BSA at Sparkbrook.
1922-1924	Light Car	Hotchkiss 1.1 ltr air-cooled V-twin, 10 hp	Two- or four-seaters. No link with Daimler.
1923-1925	14hp	1.5 litre 4cyl. Daimler sleeve valve	Various bodies. Daimler built cars.
1923-1926	16hp	1.9 litre 6-cyl. Daimler sleeve valve.	Various bodies. Daimler built cars.
1929-1936	BSA	Hotchkiss 1,021cc V-twin	Three-wheeler. No link with Daimler. Four-wheeled version built in 1932.
1933-1936	BSA	4-cylinder 1,075cc unit	Three-wheeler. No link with Daimler. Four-wheeled version built in 1933.
1933-1936	Ten	4-cyl. 1,185cc, side-valve	Lanchester Ten-based. Four- and six-light bodies.
1935-1936	Light Six	6-cyl. 1,378cc, ohv, 12 hp	Also Lanchester-based. Various sports bodies.
1935-1937	Scout	4-cylinder 1,075cc unit	Fwd. Built by BSA Cycles, no link with Daimler.

Appendix X - Royal Cars

A listing of all the Daimler and Lanchester cars used by the Royal Household.

Delivery Date	Chassis Type	Chassis No.	Body Type	Body No.
King Edward VII (HRH Prince of Wales)				
Mar. 1900	Daimler 6 hp		Mail Phaeton - Hooper	
Dec. 1900	Daimler 12 hp		Wagonette - Hooper	
Dec. 1900	Daimler 12 hp		Beaters Car - Hooper	
June. 1902	Daimler 22 hp		Tourer - Hooper	
Feb. 1903	Daimler 22 hp		Wagonette - Daimler	
July. 1904	Daimler 18/22 hp		Limousine - Hooper	
Dec. 1905	Daimler 35 hp		Limousine - Daimler	
Sept. 1907	Daimler 42 hp	4270	Shooting Brake - Hooper	4330
Oct. 1907	Daimler 35 hp		Landaulette - Daimler	
May. 1908	Daimler 58 hp	4570	Limou-Landau - Hooper	4341
King George V (HRH Prince of Wales)				
Feb. 1903	Daimler 22 hp		Tourer by Daimler	
July 1903	Daimler 22 hp		Limousine by Hooper	
1904	Daimler 28/36 hp		Limousine by Hooper	
May 1906	Daimler 30 hp		Brougham by Hooper	
May 1906	Daimler 35 hp		Limousine by Hooper	
1906	Daimler 30 hp		Shooting Brake by Daimler	
1909	Daimler 38 hp	TD6314	Limousine by Daimler	
Oct. 1909	Daimler 38 hp	TD6312	Limousine by Hooper	
1909	Daimler 57 hp	TC5074	Shooting Brake by Hooper	
June 1910	Daimler 57 hp		Limousine by Hooper	4643
Dec. 1910	Daimler 57 hp	TC5641	Limousine by Hooper	4665
Aug. 1912	Daimler 58 hp	4570	Shooting Brake by Hooper	4890
Nov. 1912	Daimler 57 hp	TG10509	Limousine by Hooper	4887
Sep. 1914	Daimler 45 hp	TB12283	Brougham by Hooper	
1920	Daimler 30 hp		Limousine by Hooper	
1923	Daimler 45 hp		Hooper body 4890 transferred	4890
Aug. 1924	Daimler 57 hp	21414	Shooting Brake by Hooper	6041
Aug. 1924	Daimler 20 hp		Limousine by Hooper	6085
Sept. 1924	Daimler 57 hp		Limousine by Hooper	5985
Sept. 1924	Daimler 57 hp		Limousine by Hooper	5998
Sept. 1924	Daimler 57 hp	21415	Shooting Brake by Hooper	6042
Aug. 1926	Daimler 35/120 hp	27046	Brougham by Hooper	6570
1928	Daimler 20/70 hp		Limousine by Daimler	
July 1929	Daimler 25/85 hp	30440	Limousine by Hooper	7208
Aug. 1929	Daimler DD6-30	30671	Brougham by Hooper	7209
Apr. 1931	Daimler 40/50 hp	32349	Limousine by Hooper	7505
Apr. 1931	Daimler 40/50 hp	32348	Limousine by Hooper	7506
Apr. 1931	Daimler 30/40 hp	32336	Limousine by Hooper	7508
May 1932	Daimler 16/20 hp	32419		
Oct. 1932	Daimler 20/25 hp	33057		
Apr. 1935	Daimler DD6-50	39701	Limousine by Hooper	8258
Aug. 1935	Daimler V26 Str.8	38438	Limousine by Hooper	8440
Queen Mary (HRH Princess of Wales)				
Aug. 1905	Daimler 28/36 hp		Limousine by H.J Mulliner	
Nov. 1906	Daimler 28/36 hp		Limou-Landau by H.J Mulliner	
Sep. 1907	Daimler 30 hp		Brougham by H.J Mulliner	
Dec. 1910	Daimler 38 hp		Limousine by Hooper	4694
Dec. 1913	Daimler 45 hp		Limousine by Hooper	5706
Oct. 1925	Daimler 45 hp	24553	Limousine by Hooper	6305
Feb. 1928	Daimler DD6-30	29440	Limousine by Hooper	6861
Apr. 1931	Daimler 40/50 hp	32347	Limousine by Hooper	7504
Apr. 1931	Daimler 30/40 hp	32335	Limousine by Hooper	7507
Dec. 1935	Daimler DD6-50	39700	Limousine by Hooper	8430
Jan. 1937	Daimler 24 hp	4372	Limousine by Hooper	8679
Nov. 1937	Daimler Str.8	44254	Limousine by Hooper	8925
July 1947	Daimler DE27	51101	Limousine by Hooper	9424
Queen Alexandra				
1910	Daimler 57 hp	TD6296	Landaulette by Hooper	
1920	Daimler 45 hp		Limou-Landau by Hooper	
1923	Daimler 45 hp		1910 Hooper body transferred	
King George VI (HRH Duke of York)				
1925	Lanchester 40 hp		Limousine	
1929	Lanchester Forty		Landaulette	
Nov. 1935	"Lanchester" Str.8	41290	Limousine by Hooper	8523
Nov. 1935	"Lanchester" Str.8	41289	Landaulette by Hooper	8524
Apr. 1937	Daimler Str.8	44209	Limousine by Hooper	8840
June 1937	Daimler Str.8	43584	Landaulette by Hooper	8857
July 1937	Daimler Str.8	44235	Shooting Brake by Hooper	8869
Aug. 1938	Daimler Str.8	48303	Landaulette by Hooper	9045
1939	Daimler EL24	47997	Limousine by Daimler	
1939	Daimler DB18	46123	Drophead coupé by Daimler	
1939	"Lanchester" Str.8	43617	Landaulette by Hooper	9114

Appendix X (continued)

A listing of all the Daimler and Lanchester cars used by the Royal Household.

Delivery Date	Chassis Type	Chassis No.	Body Type	Body No.
1940	"Lanchester" Str.8	43616	Limousine by Hooper	9164
Sept. 1941	Daimler Str.8	48314	Armoured Limou. by Hooper	9103
Sept. 1941	Daimler Str.8	48327	Armoured Limou. by Hooper	9105
May 1947	Daimler DE36	51172	Landaulette by Hooper	9336
May 1947	Daimler DE36	51177	Landaulette by Hooper	9335
June 1949	Daimler DE36	51740	Landaulette by Hooper	9541
June 1949	Daimler DE36	51741	Landaulette by Hooper	9542
July 1950	Daimler DB18SS	53806	Drophead coupe by Hooper	9617
King Edward VIII (HRH Prince of Wales, and Duke of Windsor)				
1914	Daimler 45 hp		Tourer by Barker	
July 1936	Daimler Str.8	41326	Limousine by Hooper	8599
1936	Daimler Str.8		American-styled Limousine	
Royal Stock				
Apr. 1949	Daimler DE36	51717	All-weather by Hooper	9425
Apr. 1949	Daimler DE36	51723	All-weather by Hooper	9426
Jan. 1957	Daimler DK400	92704	Landaulette by Hooper	10093
July 1956	Daimler DK400	92707	Limousine by Hooper	10094
HM Queen Elizabeth the Queen Mother				
June 1949	Daimler DE36	51732	Landaulette by Hooper	9515
Oct. 1955	Daimler DK400	92702	Limousine by Hooper	10135
Dec. 1970	Daimler DS420	1M1559	Limousine by Vanden Plas	4M1496
Aug. 1978	Daimler DS420	10031	Limousine by Vanden Plas	4M3351
Sept. 1983	Daimler DS420	200453	Limousine by Jaguar	001468
Dec. 1986	Daimler DS420	200905	Limousine by Jaguar	001918
Aug. 1992	Daimler DS420	201629	Limousine by Jaguar	002639
Queen Elizabeth II (HRH Princess Elizabeth)				
1944	Daimler DB18		Saloon by Daimler	
Feb. 1948	Daimler DE27	51040	Limousine by Hooper	9181
July 1952	Daimler DF302	80002	Empress II by Hooper	9873
1953	Daimler Conquest	83092	Saloon by Daimler	
June 1984	Daimler DS420	200477	Limousine by Jaguar	001479
July 1987	Daimler DS420	200970	Limousine by Jaguar	002003
Aug. 1988	Daimler DS420	201127	Limousine by Jaguar	002128
Sept. 1992	Daimler DS420	201630	Limousine by Jaguar	002640

Appendix XI - Daimler Commercials

This appendix lists all Daimler chassis types not related to the passenger cars. The engine manufacturer is shown in brackets, and the date refers to the types introduction.

Buses

1898 J.S. Critchley's 12 hp "Greek" omnibus (Daimler)
1899 Sidney Straker's Charabanc (Daimler)
1908 Petrol-electric double-decker (Daimler)
1910 KPL (Daimler-Knight)
1911 CC (Daimler)
1912 CB, CD (both Daimler-engined)
1914 Type Y (Daimler)
1919 CK (Daimler)
1925 CL, CM (both Daimler-engined)
1926 409, 415, 416, 417, 422, 802 (A suffix = AEC, D suffix = Daimler)
1927 423, 424, 425, 426, 427 (A suffix = AEC, D suffix = Daimler)
1929 CF6 (Daimler)
1930 CG6 (Daimler)
1931 CH6 (Daimler)
1931 CP6 (Daimler)
1934 COG5 (Gardner)
1934 COA6 (AEC)
1934 COT4 (Tangye)
1935 COG6 (Gardner)
1935 COS4 (Armstrong-Saurer)
1936 COG5/40 (Gardner)
1939 COG5/60, COG6/40 (both Gardner)
1942 CWG5 (Gardner)
1943 CWA6 (AEC)
1942 CWG5 (Gardner)
1943 CWA6 (AEC)
1946 CVA6 (AEC), CVD6 (Daimler), CVG5, CVG6 (both Gardner)
1948 CD650 (Daimler)
1949 CVG6/6 (Gardner)
1951 Freeline (Daimler or Gardner)
1952 CLD6 (Daimler), CLG5, CLG6 (both Gardner)
1956 CVG6-30 (Gardner), CVD650-30 (Daimler)
1958 CVG6LX-30, CVG6-220, CSG5, CSG6, CSG6-30 (all Gardner)
CSD6, CVD650-220 (both Daimler)
1960 Fleetline RE30 (Daimler)
1961 CGR6LW, CRG6LX (both Gardner)
1962 Roadliner SRC6 (Cummins), Roadliner SRP8 (Perkins)
1963 CCG5, CCG6 (both Guy)
1967 CVG6LW-34, CRG6LX-33, SRG6LX-33 (all Guy), CVU6LX (Gardner)
1968 SRG6LX-36, SRG6LW (both Guy), CRC6/36 (Cummins)
1970 Fleetline CRL6 (Leyland)

Trolleybuses

1936 CTC (Crompton Parkinson), CTE (English Electric), CTM (Metrovick)
(The suffix "4" or "6" referred to the number of wheels on the vehicle)

General Commercials

1899 Post Office Van (Daimler)
1908 Renard Road-Train (Daimler
1911 Type HA.57 Daimler-Foster Tractor (Daimler)
1911 Type HA.105 Daimler-Foster Tractor (Daimler)
1913 20 hp One-tonner (Daimler)
1914 Two-, Three-, Four-, and Five-Ton chassis, Ambulance (all Daimler)
1950 DC27 Ambulance (Daimler)
1962 Prototype Truck (Daimler)

Military Vehicles

1939 Scout Car (Daimler)
1941 Armoured Car (Daimler)
1944 The "Coventry" Armoured Car (Hercules)
1951 Ferret (Rolls-Royce)
1966 Fox (Jaguar)
1970 Vixen (Jaguar)

Railway Vehicles

1908 Petrol-engined Railcoach (Daimler)
1913 Petrol-electric Railcoach (Daimler)

Note: The last Daimler-engined Daimler bus was produced in 1962. From 1973, all buses were Leyland-built, and carried new designations.

Left: *The flamboyant Sir Bernard and Lady Norah Docker.*

INDEX

NB. Daimler Co. and its products mentioned throughout.